AF327512

Square-Wave Voltammetry

Valentin Mirčeski
Šebojka Komorsky-Lovrić
Milivoj Lovrić

Square-Wave Voltammetry

Theory and Application

 Springer

Šebojka Komorsky-Lovrić
Ruder Bošković Institute
Center for Marine
and Environmental Research
P.O. Box 180
10002 Zagreb
Croatia
slovric@rudjer.irb.hr

Milivoj Lovrić
Ruder Bošković Institute
Center for Marine
and Environmental Research
Bijenicka 54
10002 Zagreb
Croatia
slovric@rudjer.irb.hr

Valentin Mirčeski
Institute of Chemistry
Faculty of Natural Sciences and Mathematics
"Ss Cyril and Methodius" University
PO Box 162
1000 Skopje
Republic of Macedonia
valentin@iunona.pmf.ukim.edu.mk

ISBN-13: 978-3-540-73739-1   e-ISBN-13: 978-3-540-73740-7

Library of Congress Control Number: 2007931853

Cover design: WMX Design GmbH, Heidelberg
Typesetting and Production: LE-TEX Jelonek, Schmidt & Vöckler GbR, Leipzig

Printed on acid-free paper

9 8 7 6 5 4 3 2 1

springer.com

Preface

This is the first monograph in a series devoted to electrochemistry. Although the market is rich in books and series on electrochemical themes, it is surprising that a number of serious topics are not available. With the series "Monographs in Electrochemistry" an attempt will be made to fill these gaps. I am very thankful to the publishing house of Springer for agreeing to publish these books, and for the great freedom given to me in choosing the topics and the most competent authors, and generally for the fantastic cooperation with the publisher. I am especially thankful to Mr. Peter W. Enders.

Square-wave voltammetry is a technique that is readily available to anyone applying modern electrochemical measuring systems. Its use can be beneficial in analytical applications as well as in fundamental studies of electrode mechanisms. Upon first glance, it seems that the analytical application of square-wave voltammetry is rather simple and does not afford a deep knowledge of the background, however, this is certainly not the case. For an optimal exploitation of the potential of square-wave voltammetry, it is essential to know how the signal is generated and how its properties depend on the kinetics and thermodynamics of the electrode processes. For a detailed analysis of electrode mechanisms, this is indispensable, of course, in any case. I am very happy that three leading experts in the field of square-wave voltammetry have agreed to write the present monograph, which in fact is the first complete book on that technique ever published in English. All three authors have a long and distinguished publishing record in electroanalysis, and especially in the theory and application of square-wave voltammetry. I hope that this monograph will make it much easier for potential users in research, industrial, and environmental laboratories, etc., to apply square-wave voltammetry for their benefit.

Fritz Scholz
– Editor of the series "Monographs in Electrochemistry" –

Contents

Chapter 1
Introduction

1.1 From Square-Wave Polarography to Modern Square-Wave Voltammetry

Square-wave voltammetry (SWV) is a powerful electrochemical technique that can be applied in both electrokinetic and analytic measurements [1–5]. The technique originates from the Kalousek commutator [6] and Barker's square-wave polarography [7]. Kalousek constructed an instrument with a rotating commutator that switched the potential of the dropping mercury electrode between two voltage levels with the frequency of five cycles per second [8,9]. Three methods for programming the voltages have been devised and designated as types I, II, and III, and these are shown in Fig. 1.1. Type I polarograms were recorded by superimposing a low-amplitude square-wave ($20-50$ mV peak-to-peak) onto the ramp voltage of conventional polarography. The current was recorded during the higher potential half-cycles only. Figure 1.2 shows the theoretical response of a simple and electrochemically reversible electrode reaction:

$$O^{m+} + ne^- \leftrightharpoons R^{(m-n)+} \tag{1.1}$$

obtained by the type I program. Only the reactant O^{m+} is initially present in the bulk of the solution. The starting potential is -0.25 V vs. $E_{1/2}$, where $E_{1/2}$ is a half-wave potential of dc polarogram of electrode reaction (1.1). The response is characterized by a maximum oxidation current appearing at 0.034 V vs. $E_{1/2}$. In the vicinity of the half-wave potential, the reactant is reduced during the lower potential half-cycle (which is not recorded) and the product is oxidized during the higher potential half-cycle, which is recorded. This is illustrated in Fig. 1.3, which shows theoretical concentrations of the reactant and product near the electrode surface at the end of the last cathodic (A) and anodic (B) half-cycles applied to the same drop. The method was improved by Ishibashi and Fujinaga, who introduced the differential polarography by measuring the difference in current between successive half-cycles of the square-wave signal [10–12]. The frequency of the signal was 14 Hz. It was superimposed on a rapidly changing potential ramp and applied to the single mercury

V. Mirčeski, Š. Komorsky-Lovrić, M. Lovrić, *Square-Wave Voltammetry*
doi: 10.1007/978-3-540-73740-7, ©Springer 2008

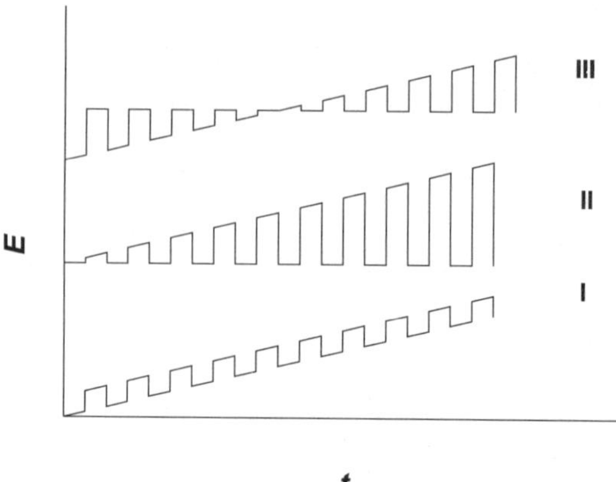

Fig. 1.1 Potential-time relationships realized by the Kalousek commutator

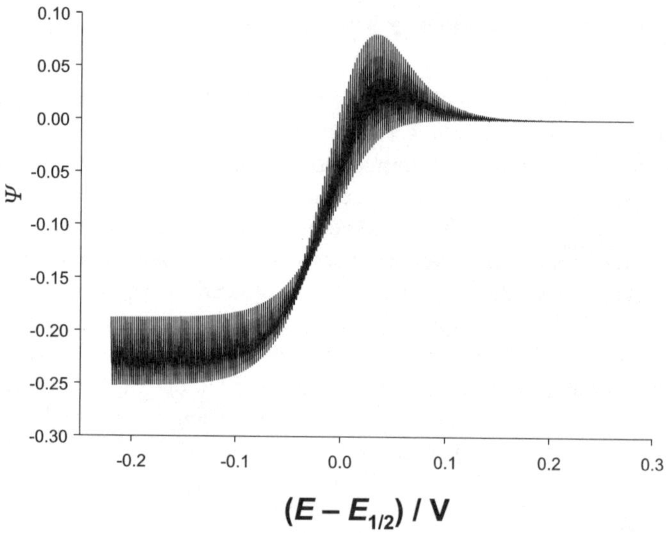

Fig. 1.2 Type I Kalousek polarogram of electrode reaction (1.1) on dropping mercury electrode. $\Psi = I/nFA_m c_O^*(Df)^{1/2}$, frequency = 5 Hz, amplitude = 30 mV, drop life time = 1 s, $dE/dt = 2$ mV/s and $E_{st} = -0.25$ V vs. $E_{1/2}$. For the meaning of symbols, see below (1.9) and (1.24)

drop. Barker and Jenkins introduced three important innovations: (i) the frequency of square-wave signal was 225 Hz, (ii) the current was measured during the last 280 µs of each half-cycle and the difference between two successive readings was recorded, and (iii) the measurement was performed only once in the life of each drop 250 ms before its end [13, 14]. Figure 1.5 shows the theoretical square-wave polarogram of the electrode reaction (1.1) under the same conditions as in Fig. 1.2. The response is a bell-shaped current-voltage curve with its maximum at -0.016 V vs. $E_{1/2}$. Each current-voltage step corresponds to a separate mercury drop. The objec-

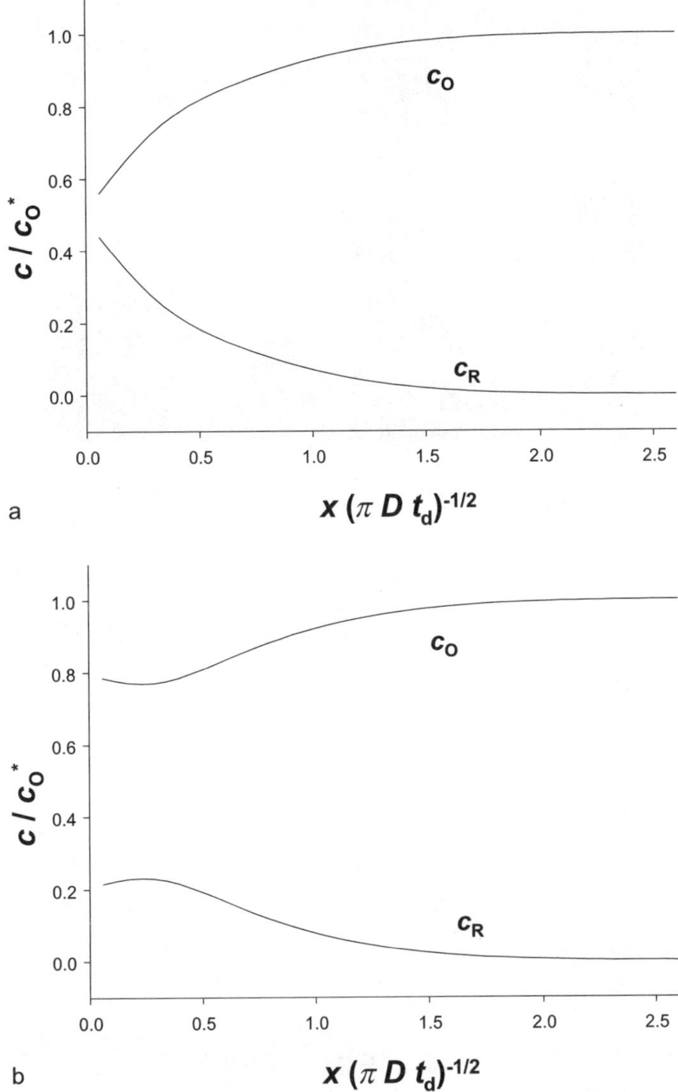

Fig. 1.3 Concentrations of the reactant and product of the electrode reaction (1.1) near the electrode surface at the end of the last cathodic (**a**) and anodic (**b**) Kalousek type I half-cycles applied to the same drop. $E = 0.004$ (**a**) and 0.034 V vs. $E_{1/2}$ (**b**). All other data are as in Fig. 1.2

tive of Barker's innovations was to minimize the influence of capacity current, i.e., to discriminate that current with respect to the faradaic current. During each half-cycle, the double layer charging current decreases exponentially with time, while the faradaic current is inversely proportional to the square-root of time. Under certain conditions, the charging current at the end of half-cycle may be smaller than the faradaic current. A theoretical example is shown in Fig. 1.6. Generally, the charg-

Fig. 1.4 Portraits of Mirko Kalousek (*left*) and Geoffrey Barker (*right*) (reprinted from [66] and [67] with permission)

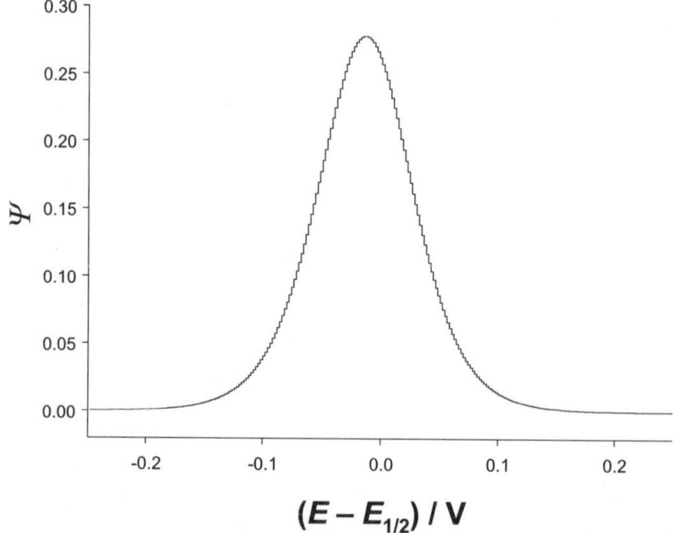

Fig. 1.5 Barker square-wave polarogram of electrode reaction (1.1) on dropping mercury electrode. $\Psi = I/nFA_m c_O^*(Df)^{1/2}$, frequency $= 225\,\text{Hz}$, amplitude $= 30\,\text{mV}$, drop life time $= 1\,\text{s}$, $dE/dt = 2\,\text{mV/s}$ and $E_{\text{st}} = -0.25\,\text{V}$ vs. $E_{1/2}$

ing current is partly eliminated by the subtraction of currents measured at the end of two successive half-cycles. This is because the charging current depends on the difference between the electrode potential and its potential of zero charge. If the square-wave amplitude is small, the difference between the charging currents of the cathodic and anodic half-cycles is also small, and for this reason, square-wave

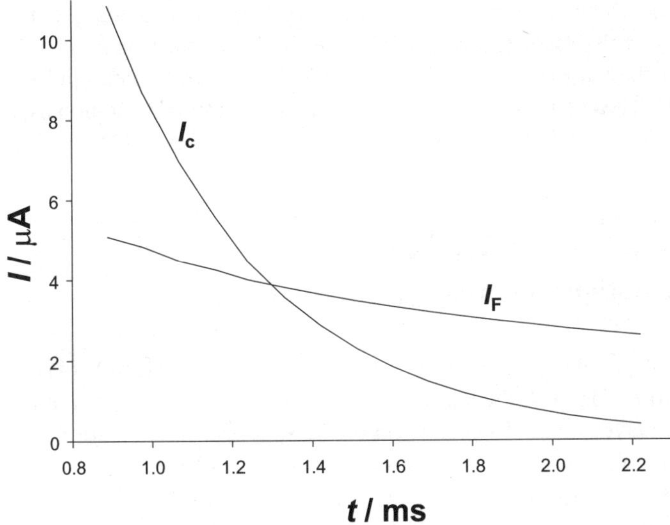

Fig. 1.6 A scheme of a double layer charging current (I_C) and the Faradaic current (I_F) during the second half of the last half-cycle of square-wave signal applied to the dropping mercury electrode. $E - E_{1/2} = -0.016$ V, $E - E_{pzc} = 0.1$ V, $C = 40\,\mu F/cm^2$, $R = 10\,\Omega$, $A_m = 0.01$ cm^2, $D = 9 \times 10^{-6}$ cm^2/s, $n = 1$, $c_O^* = 5 \times 10^{-4}$ mol/L and $f = 225$ Hz

voltammetry and differential pulse polarography and voltammetry are discriminating against a capacitive current [1, 3, 15–19].

This method was developed further by superimposing the square-wave signal onto a staircase signal [20, 21]. Some of the possible potential-time waveforms are shown in Fig. 1.7. Usually, each square-wave cycle occurs during one stair-

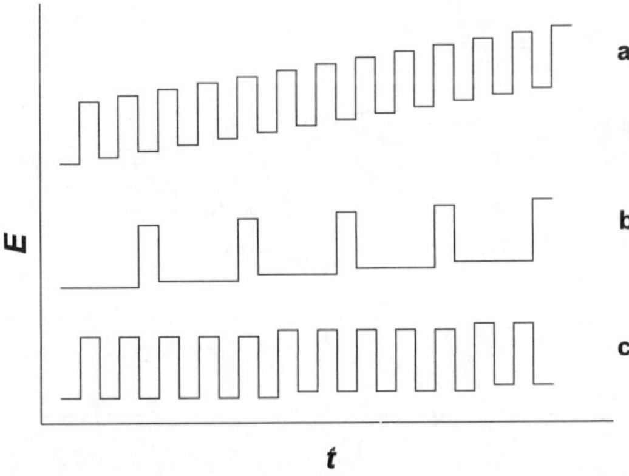

Fig. 1.7 Potential-time waveforms obtained by superimposing the square-wave signal onto a staircase signal: square-wave voltammetry (*a*), differential pulse voltammetry (*b*) and multiple square-wave voltammetry (*c*)

case period, which is sometimes called Osteryoung SWV [20–23], but in multiple square-wave voltammetry, several cycles are applied on the single tread [24,25]. The asymmetric signal (b) in Fig. 1.7 is a general form of differential pulse voltamme-try [22, 23]. These complex excitation signals were applied to stationary electrodes, or a single mercury drop. More details can be found in several reviews [26–40].

1.2 Square-Wave Voltammetry: Calculations and Instrumentation

Figure 1.8 shows the potential-time waveform of the modern SWV [41]. Comparing to curve (a) in Fig. 1.7, the starting potential is a median of extreme potentials of the square-wave signal. To each tread of the staircase signal a single square-wave cycle is superimposed, so the waveform can be considered as a train of pulses to-wards higher and lower potentials added to the potential that changes in a stepwise manner. The magnitude of each pulse, E_{sw}, is one-half of the peak-to-peak ampli-tude of the square-wave signal. For historical reasons, the pulse height E_{sw} is called the square-wave amplitude [1]. The duration of each pulse is one-half the staircase period: $t_p = \tau/2$. The frequency of the signal is the reciprocal of the staircase period: $f = 1/\tau$. The potential increment ΔE is the height of the staircase waveform. Re-lative to the scan direction, ΔE, forward and backward pulses can be distinguished. The currents are measured during the last few microseconds of each pulse and the difference between the current measured on two successive pulses of the same step

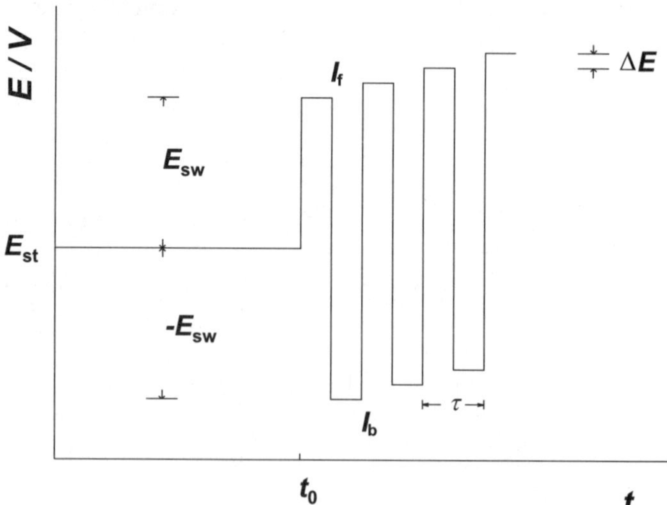

Fig. 1.8 Scheme of the square-wave voltammetric excitation signal. E_{st} starting potential, E_{sw} pulse height, ΔE potential increment, τ staircase period, t_0 delay time and I_f and I_b denote the points where the forward and backward currents are sampled, respectively

is recorded as a net response ($\Delta I = I_f - I_b$). For analytical purposes, several instantaneous currents can be sampled at certain intervals during the last third, or some other portion of the pulse, and then averaged. This is done because the response appears less noisy if the sampling window is wider [42,43]. The two components of the net response, I_f and I_b, i.e., the currents of the forward and the backward series of pulses, respectively, are also displayed. The currents are plotted as a function of the corresponding potential of the staircase waveform. The delay period, t_0, which may precede the signal, is used for the accumulation of the reactant on the working electrode surface in order to record the stripping process.

SWV experiments are usually performed on stationary solid electrodes or static mercury drop electrodes. The response consists of discrete current-potential points separated by the potential increment ΔE [1, 20–23]. Hence, ΔE determines the apparent scan rate, which is defined as $\Delta E/\tau$, and the density of information in the response, which is a number of current-potential points within a certain potential range. The currents increase proportionally to the apparent scan rate. For better graphical presentation, the points can be interconnected, but the line between two points has no physical significance, as there is no theoretical reason to interpolate any mathematical function between two experimentally determined current-potential points. The currents measured with smaller ΔE are smaller than the values predicted by the interpolation between two points measured with bigger ΔE [3]. Frequently, the response is distorted by electronic noise and a smoothing procedure is necessary for its correct interpretation. In this case, it is better if ΔE is as small as possible. By smoothing, the set of discrete points is transformed into a continuous current-potential curve. Care should be taken that the smoothing procedure does not distort the square-wave response.

A solution of the diffusion equation for an electrode reaction for repetitive stepwise changes in potential can be obtained by numerical integration [44]. For a stationary planar diffusion model of a simple, fast, and reversible electrode reaction (1.1), the following differential equations and boundary conditions can be formulated:

$$\partial c_O/\partial t = D(\partial^2 c_O/\partial x^2) \tag{1.2}$$

$$\partial c_R/\partial t = D(\partial^2 c_R/\partial x^2) \tag{1.3}$$

Initially, only the reactant O^{m+} is present in the solution and its concentration is uniform:

$$t = 0: \quad c_O = c_O^*, \; c_R = 0 \tag{1.4}$$

At the infinite distance from the electrode surface, the concentrations of the reactant and product do not change:

$$t > 0, \, x \to \infty: \quad c_O \to c_O^*, \; c_R \to 0 \tag{1.5}$$

The current is proportional to the gradient of concentration of product at the electrode surface:

$$x = 0: \quad D(\partial c_O/\partial x)_{x=0} = -I/nFA \tag{1.6}$$

$$D(\partial c_R/\partial x)_{x=0} = I/nFA \tag{1.7}$$

The concentrations of reactant and product at the electrode surface are connected by the Nernst equation:

$$(c_O)_{x=0} = (c_R)_{x=0} \exp(\varphi) \tag{1.8}$$

$$\varphi = (nF/RT)(E - E^\theta) \tag{1.9}$$

Here c_O and c_R are the concentrations of the reactant and product, respectively, D is a common diffusion coefficient, c_O^* is the bulk concentration of the reactant, I is the current, n is the number of electrons, F is the Faraday constant, A is the electrode surface area, E is electrode potential, E^θ is the standard potential, R is the gas constant, T is absolute temperature, x is the distance from the electrode surface and t is the time variable [45].

Using the Laplace transformations of the reactant concentration and its derivative on time [46]:

$$\mathscr{L}c_O = \int_0^\infty c_O \exp(-st)\, dt \tag{1.10}$$

$$\mathscr{L}(\partial c_O/\partial t) = s\mathscr{L}c_O - (c_O)_{t=0} \tag{1.11}$$

the differential equation (1.2) can be transformed into:

$$\partial^2 u/\partial x^2 - (s/D)\, u = 0 \tag{1.12}$$

where $u = \mathscr{L}c_O - c_O^*/s$ and s is the transformation variable. The boundary conditions are:

$$x \rightarrow \infty: \quad u \rightarrow 0 \tag{1.13}$$

$$x = 0: \quad (\partial u/\partial x)_{x=0} = -\mathscr{L}I/nFAD \tag{1.14}$$

The general solution of (1.12) is:

$$u_{1,2} = K_1 \exp(-(s/D)^{1/2}x) + K_2 \exp((s/D)^{1/2}x) \tag{1.15}$$

A particular solution is obtained by using (1.13) and (1.14):

$$K_2 = 0 \tag{1.16}$$

$$K_1 = s^{-1/2}\mathscr{L}I/nFAD^{1/2} \tag{1.17}$$

$$(\mathscr{L}c_O)_{x=0} = c_O^*/s + s^{-1/2}\mathscr{L}I/nFAD^{1/2} \tag{1.18}$$

By the inverse Laplace transformation of (1.18) an integral equation is obtained [46]:

$$(c_O)_{x=0} = c_O^* + (nFA)^{-1}(D\pi)^{-1/2} \int_0^t I(\tau)(t-\tau)^{-1/2}\,d\tau \qquad (1.19)$$

Within each time interval from 0 to t, the current I depends on the variable τ. The integral $\int f(\tau)g(t-\tau)\,d\tau$ is called the convolution of functions f and g.

The solution of (1.3) is obtained by the same procedure:

$$(c_R)_{x=0} = -(nFA)^{-1}(D\pi)^{-1/2} \int_0^t I(\tau)(t-\tau)^{-1/2}\,d\tau \qquad (1.20)$$

The convolution integral in (1.19) and (1.20) can be solved by the method of numerical integration proposed by Nicholson and Olmstead [47]. The time t is divided into m time increments: $t = md$. It is assumed that within each time increment the function I can be replaced by the average value I_j:

$$\int_0^t I(\tau)(t-\tau)^{-1/2}\,d\tau = \sum_{j=1}^m I_j \int_{(j-1)d}^{jd} (md-\tau)^{-1/2}\,d\tau \qquad (1.21)$$

The integral in (1.21) is solved by the substitution $p = md - \tau$:

$$\int_{(j-1)d}^{jd} (md-\tau)^{-1/2}\,d\tau = 2d^{1/2}[(m-j+1)^{1/2} - (m-j)^{1/2}] \qquad (1.22)$$

Each square-wave half-period is divided into 25 time increments: $d = (50f)^{-1}$. By introducing (1.19) and (1.20) into (1.8), the following system of recursive formulae is obtained:

$$\Psi_1 = -5(\pi/2)^{1/2}(1+\exp(\varphi_1))^{-1} \qquad (1.23)$$

$$\Psi_m = -5(\pi/2)^{1/2}(1+\exp(\varphi_m))^{-1} - \sum_{j=1}^{m-1} \Psi_j S_{m-j+1} \qquad (1.24)$$

where $\Psi = I/nFAc_O^*(Df)^{1/2}$, $S_1 = 1$, $S_k = k^{1/2} - (k-1)^{1/2}$, $\varphi_m = (nF/RT)(E_m - E^\theta)$, $m = 2, 3, \ldots M$ and $M = -50\,(E_{fin} - E_{st})/\Delta E$. The potential E_m changes according to Fig. 1.8.

If the electrode reaction (1.1) is kinetically controlled, (1.8) must be substituted by the Butler–Volmer equation:

$$I/nFA = -k_s \exp(-\alpha\varphi)[(c_O)_{x=0} - (c_R)_{x=0}\exp(\varphi)] \qquad (1.25)$$

where k_s is the standard rate constant and α is the cathodic transfer coefficient. In this case, the following recursive formulae are obtained [44, 48–50]:

$$\Psi_1 = -\frac{\kappa \exp(-\alpha \varphi_1)}{1 + \kappa \exp(-\alpha \varphi_1) \frac{\sqrt{2}}{5\sqrt{\pi}} [1 + \exp(\varphi_1)]} \tag{1.26}$$

$$\Psi_m = -Z_1 - Z_2 \sum_{j=1}^{m-1} \Psi_j S_{m-j+1} \tag{1.27}$$

$$Z_1 = \frac{\kappa \exp(-\alpha \varphi_m)}{1 + \kappa \exp(-\alpha \varphi_m) \frac{\sqrt{2}}{5\sqrt{\pi}} [1 + \exp(\varphi_m)]} \tag{1.28}$$

$$Z_2 = \frac{\kappa \exp(-\alpha \varphi_m) \frac{\sqrt{2}}{5\sqrt{\pi}} [1 + \exp(\varphi_m)]}{1 + \kappa \exp(-\alpha \varphi_m) \frac{\sqrt{2}}{5\sqrt{\pi}} [1 + \exp(\varphi_m)]} \tag{1.29}$$

where $\kappa = k_s (Df)^{-1/2}$ is a dimensionless kinetic parameter and the meanings of other symbols are given below (1.24).

The developments of instrumentation for Kalousek [31, 51, 52] and square-wave polarography [53–58] and square-wave voltammetry [59–65] are mainly of historical interest. Today, many computer-controlled potentiostats/galvanostats providing SWV signal generation are available from numerous manufacturers, such as Eco-Chemie (models PGSTAT 10, 12, 20, 30, 100 and 302 and μAutolab I, II, and III), Metrohm (models VA 646 and 797 Computrace), Princeton Applied Research (models 263A, 273A, 283, 2263, 2273, and 384B), Bioanalytical Systems (models 100A, 100B/W and Epsilon), Cypress Systems (models CS 1090 and 1200 and CYSY-2), Amel Srl (models 433, 7050 and 7060), Gamry Instruments (models PCI 4/300 and 4/750), Analytical Instrument Systems (models LCP-200 and DLK-100), Uniscan Instruments (model PG 580), Palm Instruments (model Palmsens), Rudolph Instruments (model GATTEA 4000 AS) and IVA Company (model IVA-5).

References

1. Osteryoung J, O'Dea JJ (1986) Square-wave voltammetry. In: Bard AJ (ed) Electroanalytical chemistry, vol 14. Marcel Dekker, New York, p 209
2. Eccles GN (1991) Crit Rev Anal Chem 22:345
3. Lovrić M (2002) Square-wave voltammetry. In: Scholz F (ed) Electroanalytical methods, Springer, Berlin Heidelberg New York, p 111
4. de Souza D, Machado SAS, Avaca LA (2003) Quim Nova 26:81
5. de Souza D, Codognoto L, Malagutti AR, Toledo RA, Pedrosa VA, Oliveira RTS, Mazo LH, Avaca LA, Machado SAS (2004) Quim Nova 27:790
6. Ružić I (1972) J Electroanal Chem 39:111
7. Barker GC, Gardner AW (1992) Analyst 117:1811
8. Kalousek M (1946) Chem Listy 40:149
9. Kalousek M (1948) Coll Czechoslov Chem Commun 13:105
10. Ishibashi M, Fujinaga T (1950) Bull Chem Soc Jpn 23:261
11. Ishibashi M, Fujinaga T (1952) Bull Chem Soc Jpn 25:68
12. Ishibashi M, Fujinaga T (1952) Bull Chem Soc Jpn 25:238
13. Barker GC, Jenkins IL (1952) Analyst 77:685
14. Barker GC (1958) Anal Chim Acta 18:118
15. Turner JA, Christie JH, Vuković M, Osteryoung RA (1977) Anal Chem 49:1904
16. Barker GC, Gardner AW (1979) J Electroanal Chem 100:641
17. Stefani S, Seeber R (1983) Ann Chim 73:611
18. Dimitrov JD (1997) Anal Lab 6:87
19. Dimitrov JD (1998) Anal Lab 7:3
20. Ramaley L, Krause MS Jr (1969) Anal Chem 41:1362
21. Krause MS Jr, Ramaley L (1969) Anal Chem 41:1365
22. Rifkin SC, Evans DH (1976) Anal Chem 48:1616
23. Christie JH, Turner JA, Osteryoung RA (1977) Anal Chem 49:1899
24. Fatouros N, Simonin JP, Chevalet J, Reeves RM (1986) J Electroanal Chem 213:1
25. Krulic D, Fatouros N, Chevalet J (1990) J Electroanal Chem 287:215
26. Milner GWC, Slee LJ (1957) Analyst 82:139
27. Hamm RE (1958) Anal Chem 30:351
28. Saito Y, Okamoto K (1962) Rev Polarog 10:227
29. Kaplan BY, Sorokovskaya II (1962) Zavod Lab 28:1053
30. Geerinck G, Hilderson H, Vanttulle C, Verbeck F (1963) J Electroanal Chem 5:48
31. Kinard WF, Philp RH, Propst RC (1967) Anal Chem 39:1556
32. Geissler M, Kuhnhardt C (1970) Square-wave polarographie. VEB Deutscher Verlag für Grundstoffindustrie, Leipzig
33. Kaplan BY, Sevastyanova TN (1971) Zh Anal Khim 26:1054
34. Igolinskii VA, Kotova NA (1973) Elektrokhimiya 9:1878

35. Sturrock PE, Carter RJ (1975) Crit Rev Anal Chem 5:201
36. Blutstein H, Bond AM (1976) Anal Chem 48:248
37. Kopanica M, Stara V (1981) J Electroanal Chem 127:255
38. Alexander PW, Akapongkul V (1984) Anal Chim Acta 166:119
39. Ramaley L, Tan WT (1981) Can J Chem 59:3326
40. Hwang JY, Wang YY, Wan CC (1986) J Chin Chem Soc 33:303
41. Osteryoung JG, Osteryoung RA (1985) Anal Chem 57:101 A
42. Aoki K, Maeda K, Osteryoung J (1989) J Electroanal Chem 272:17
43. Zachowski EJ, Wojciechowski M, Osteryoung J (1986) Anal Chim Acta 183:47
44. O'Dea JJ, Osteryoung J, Osteryoung RA (1981) Anal Chem 53:695
45. Galus Z (1994) Fundamentals of electrochemical analysis. Ellis Horwood, New York, Polish
    Scientific Publishers PWN, Warsaw
46. Spiegel MR (1965) Theory and problems of Laplace transforms. McGraw-Hill, New York
47. Nicholson RS, Olmstead ML (1972) Numerical solutions of integral equations. In: Matson JS,
    Mark HB, MacDonald HC (eds) Electrochemistry: calculations, simulations and instrumenta-
    tion, vol 2. Marcel Dekker, New York, p 119
48. O'Dea JJ, Osteryoung J, Osteryoung RA (1983) J Phys Chem 87:3911
49. O'Dea JJ, Osteryoung J, Lane T (1986) J Phys Chem 90:2761
50. Nuwer MJ, O'Dea JJ, Osteryoung J (1991) Anal Chim Acta 251:13
51. Radej J, Ružić I, Konrad D, Branica M (1973) J Electroanal Chem 46:261
52. Paspaleev E, Batsalova K, Kunchev K (1979) Zavod Lab 45:504
53. Ferrett DJ, Milner GWC, Shalgosky HI, Slee LJ (1956) Analyst 81:506
54. Taylor JH (1964) J Electroanal Chem 7:206
55. Buchanan EB, McCarten JB (1965) Anal Chem 37:29
56. Kalvoda R, Holub I (1973) Chem Listy 67:302
57. Barker GC, Gardner AW, Williams MJ (1973) J Electroanal Chem 42:App 21
58. Buchanan EB Jr, Sheleski WJ (1980) Talanta 27:955
59. Yarnitzky C, Osteryoung RA, Osteryoung J (1980) Anal Chem 52:1174
60. He P, Avery JP, Faulkner LR (1982) Anal Chem 54:1313 A
61. Anderson JE, Bond AM (1983) Anal Chem 55:1934
62. Lavy-Feder A, Yarnitzky C (1984) Anal Chem 56:678
63. Jayaweera P, Ramaley L (1986) Anal Instrum 15:259
64. Poojary A, Rajagopalan SR (1986) Indian J Technol 24:501
65. Wong KH, Osteryoung RA (1987) Electrochim Acta 32:629
66. Bard AJ, Inzelt G, Scholz F (2007) Electrochemical dictionary. Springer, Berlin Heidelberg
    New York
67. Parsons R (1977) J Electroanal Chem 75:1

Chapter 2
Electrode Mechanisms

2.1 Electrode Reactions of Dissolved Species on Stationary Planar Electrodes

2.1.1 Reversible Electrode Reactions

Figure 2.1 shows computed square-wave voltammogram of the simple, fast and electrochemically reversible electrode reaction (1.1), i.e., $O^{m+} + ne^- \leftrightarrows R^{(m-n)+}$. The response was calculated by using (1.23) and (1.24). The dimensionless net response $(\Delta\Psi = -\Delta I/nFAc_O^*(Df)^{1/2})$, where $\Delta I = I_f - I_b$, and its forward (reductive) (Ψ_f), and backward (oxidative) (Ψ_b) components are shown. The meanings of other symbols are given below (1.9). The voltammogram is characterized by the maximum net response, which is also called the net peak current, ΔI_p. The corresponding staircase potential is the net peak potential E_p. Other characteristics are the minimum of the reductive component, the maximum of the oxidative component and their potentials. The net peak potential and the peak potentials of both components are independent of SW frequency. This is one of various indications that the electrode reaction is electrochemically reversible within the range of applied frequencies [1–6].

Both the dimensionless net peak current $\Delta\Psi_p$ and the peak width at half-height, or the half-peak width, $\Delta E_{p/2}$ depend on the products "the number of electrons times the SW amplitude", i.e., nE_{sw}, and "the number of electrons times the potential increment", i.e., $n\Delta E$. This is shown in Table 2.1 and Fig. 2.2 (curve 1), for the constant product $n\Delta E$. With increasing nE_{sw} the slope $\partial\Delta\Psi_p/\partial nE_{sw}$ continuously decreases, while the gradient $\partial\Delta E_{p/2}/\partial nE_{sw}$ increases. The maximum ratio $\Delta\Psi_p/\Delta E_{p/2}$ appears for $nE_{sw} = 50\,\text{mV}$, as can be seen in Fig. 2.2. This is the optimum SW amplitude for analytical measurement [7]. At higher amplitudes the resolution of two peaks is diminished. The ratio of peak currents of the forward and backward components, and the peak potentials of the components are also listed in Table 2.1. If $nE_{sw} > 10\,\text{mV}$, the backward component indicates the reversibility of the electrode reaction, and if $nE_{sw} > 5\,\text{mV}$, the net peak potential E_p is equal to the half-wave potential of the reversible reaction (1.1). If $E_{sw} = 0$, the square-wave sig-

V. Mirčeski, Š. Komorsky-Lovrić, M. Lovrić, *Square-Wave Voltammetry*   13
doi: 10.1007/978-3-540-73740-7, ©Springer 2008

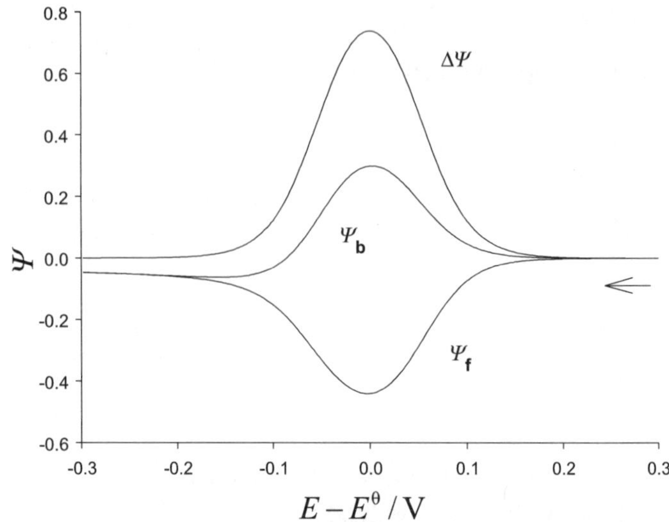

Fig. 2.1 Theoretical square-wave voltammogram of fast and reversible electrode reaction (1.1). $E_{sw} = 50$ mV, $n = 1$, $E_{st} - E^{\theta} = 0.3$ V, $t_0 = 0$ s and $\Delta E = -2$ mV. The dimensionless net response ($\Delta\Psi$) and its forward (Ψ_f) and backward (Ψ_b) components

Table 2.1 Square-wave voltammetry of fast and reversible electrode reaction (1.1). The dimensionless net peak current, the ratio of peak currents of the forward and backward components, the peak potentials of the components and the half-peak width as functions of SW amplitude; $n\Delta E = -2$ mV

$nE_{sw}/$mV	$\Delta\Psi_p$	$I_{p,f}/I_{p,b}$	$E_{p,f} - E_p/$mV	$E_{p,b} - E_p/$mV	$\Delta E_{p/2}/$mV
10	0.1961	-10.35	-12	26	92
20	0.3701	-2.78	-8	10	96
30	0.5206	-1.94	-6	6	104
40	0.6432	-1.63	-4	4	112
50	0.7383	-1.47	-2	2	124
60	0.8093	-1.37	-2	2	139
70	0.8608	-1.31	0	0	152
80	0.8975	-1.27	2	-2	168
90	0.9231	-1.23	4	-2	186
100	0.9409	-1.21	6	-4	204

nal turns into the signal of differential staircase voltammetry [8–10], and $\Delta\Psi_p$ does not vanish [4]. To establish an additional criterion of the reversibility of the reaction (1.1), the standard SW amplitudes $E_{sw} = 50$, 25 and 15 mV, for $n = 1$, 2 and 3, respectively, and the common potential increment $\Delta E = -2$ mV are proposed. The characteristic data of responses of simple and electrochemically reversible electrode reactions under standard conditions are listed in Table 2.2.

The net peak current depends linearly on the square root of the frequency [5, 11]:

$$\Delta I_p = -nFAD^{1/2}\Delta\Psi_p f^{1/2} c_O^* \tag{2.1}$$

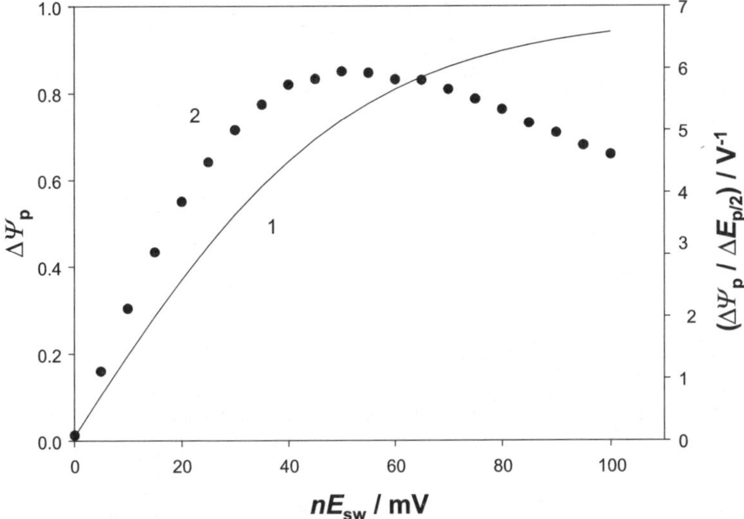

Fig. 2.2 The dependence of the dimensionless net peak current (*1*) and the ratio of the dimensionless net peak current and the half-peak width (*2*) on the product of number of electrons and the square-wave amplitude

Table 2.2 Criteria of the reversibility of reaction (1.1). Properties of the response under standard conditions

n	$\Delta E/\mathrm{mV}$	$E_{\mathrm{sw}}/\mathrm{mV}$	$I_{\mathrm{p,f}}/I_{\mathrm{p,b}}$	$E_{\mathrm{p,f}} - E_{\mathrm{p}}/\mathrm{V}$	$E_{\mathrm{p,b}} - E_{\mathrm{p}}/\mathrm{V}$	$\Delta E_{\mathrm{p/2}}/\mathrm{mV}$
1	-2	50	-1.47	-0.002	0.002	124
2	-2	25	-1.70	-0.002	0.002	62
3	-2	15	-2.03	-0.002	0.002	40

The condition is that the instantaneous current is sampled during the last few microseconds of the pulse [2, 3]. This procedure was assumed in the theoretical calculations presented in Figs. 2.1 and 2.2, and Tables 2.1 and 2.2. Usually, the current is sampled during a certain portion of the pulse and then averaged. The average response corresponds to an instantaneous current sampled in the middle of the sampling window. For $nE_{\mathrm{sw}} = 25\,\mathrm{mV}$ and $n\Delta E = -5\,\mathrm{mV}$, this relationship is [12]:

$$\Delta I_{\mathrm{p}} = -nFAD^{1/2}c_{\mathrm{O}}^{*}\,0.477t_{\mathrm{s}}^{-1/2}(1 - 0.317\beta^{1/2}) \tag{2.2}$$

where t_{s} is the sampling time and $\beta = t_{\mathrm{s}}/t_{\mathrm{p}}$ is the pulse fraction at which the current is sampled (see Fig. 2.3). If β is constant, the relationship between ΔI_{p} and the square-root of the time variable is linear, regardless of whether the frequency, or the reciprocal of sampling time is used as the characteristic variable. This condition is satisfied if the relative size of the sampling window (s/t_{p}) is constant, because $\beta = 1 - s/2t_{\mathrm{p}}$. If the absolute size of the sampling window is constant (e.g., $s = 1\,\mathrm{ms}$), its

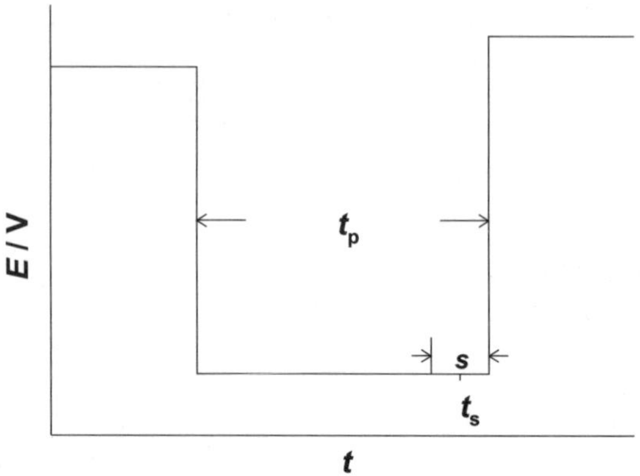

Fig. 2.3 A scheme of the current sampling procedure in pulse techniques

relative size increases and β decreases as the frequency is increased. So, the product $\Delta I_p t_s^{1/2}$ increases with the increasing frequency.

The second condition is that there is no significant uncompensated resistance in the electrochemical cell [13, 14]. The influence of IR-drop increases with SW frequency, the ratio $\Delta I_p / f^{1/2}$ decreases with frequency and the net peak current is not a linear function of the square-root of frequency. A theoretical example is shown in Fig. 2.4.

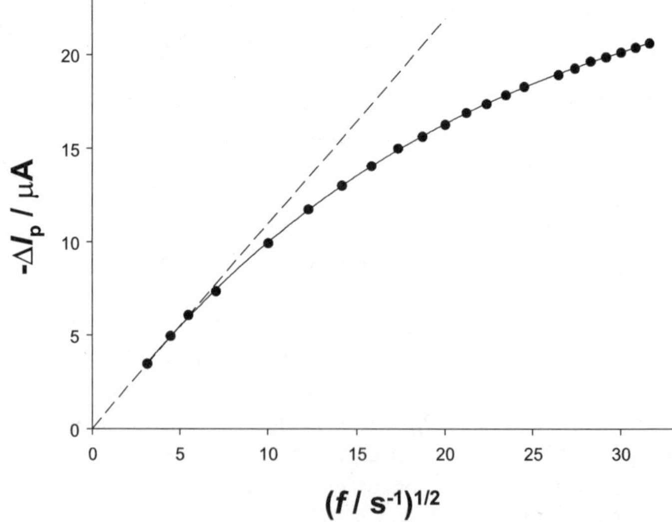

Fig. 2.4 Theoretical dependence of the net peak current of reversible reaction (1.1) on the square-root of SW frequency under the influence of uncompensated resistance in the cell. The *dotted line* is the expected linear relationship in the absence of resistance. $A = 0.0147\,\text{cm}^2$, $n = 1$, $D = 5 \times 10^{-6}\,\text{cm}^2/\text{s}$, $c_O^* = 1 \times 10^{-3}\,\text{mol/L}$, $R = 1.14\,\text{k}\Omega$, $E_{sw} = 20\,\text{mV}$ and $\Delta E = -5\,\text{mV}$

2.1.2 Kinetically Controlled Electrode Reaction

If the reaction (1.1) is controlled by the electrode kinetics, i.e., when the electrode reaction is not electrochemically reversible, the response depends on the dimensionless kinetic parameter $\kappa = k_s(Df)^{-1/2}$ and the transfer coefficient α [15–17]. Typical voltammograms are shown in Figs. 2.5 and 2.6. They were calculated using (1.26)–(1.29). Figure 2.7 shows the dependence of the dimensionless net peak current $\Delta\Psi_p$ on the logarithm of kinetic parameter κ. The reaction is reversible if

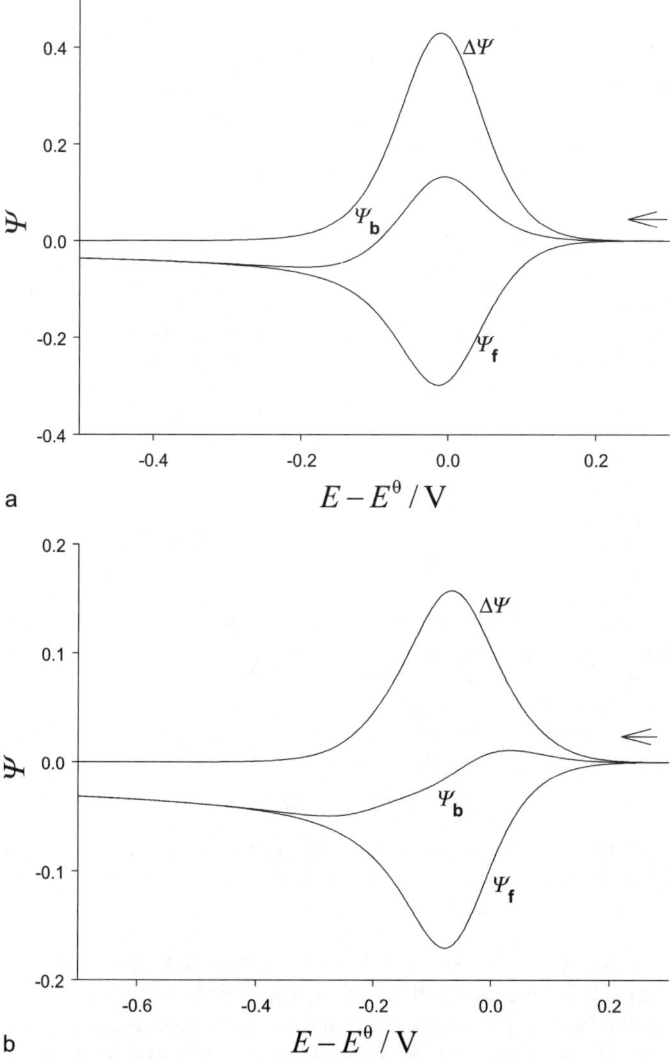

Fig. 2.5a,b Square-wave voltammogram of quasireversible electrode reaction (1.1); $\kappa = 0.3$ (**a**) and 0.05 (**b**), $\alpha = 0.5$, $nE_{sw} = 50\,\text{mV}$, $n\Delta E = -2\,\text{mV}$, $t_0 = 0\,\text{s}$ and $E_{st} - E^\theta = 0.3\,\text{V}$

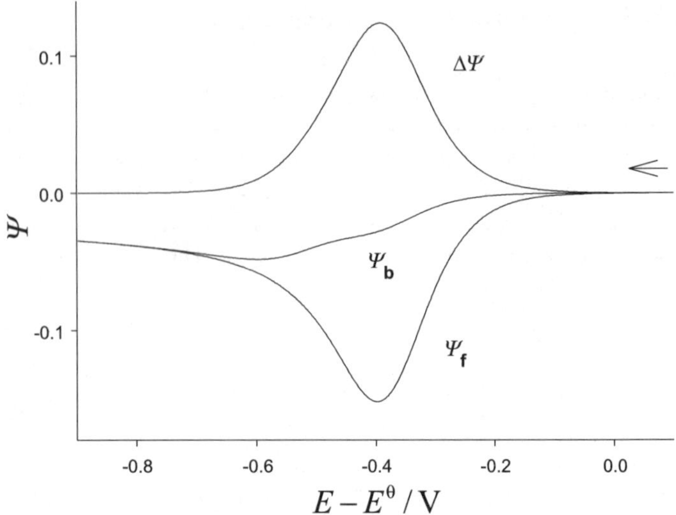

Fig. 2.6 SWV of irreversible electrode reaction (1.1); $\kappa = 10^{-4}$, $\alpha = 0.5$, $nE_{sw} = 50\,\mathrm{mV}$, $n\Delta E = -2\,\mathrm{mV}$, $t_0 = 0\,\mathrm{s}$ and $E_{st} - E^\theta = 0.3\,\mathrm{V}$

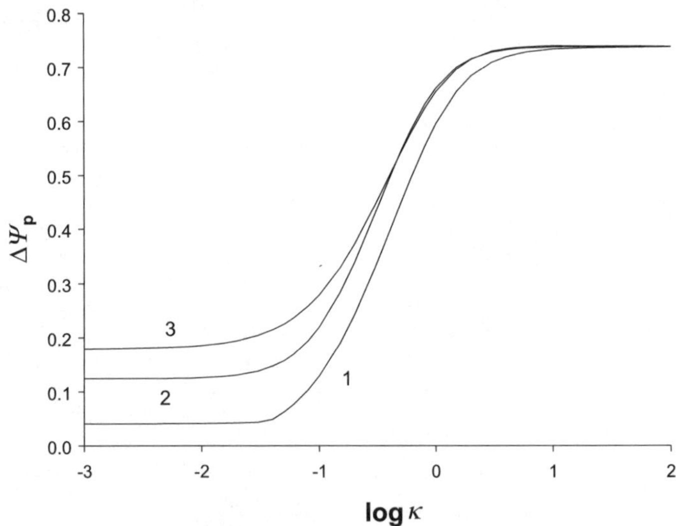

Fig. 2.7 The dependence of the dimensionless net peak current on the logarithm of the dimensionless kinetic parameter; $\alpha = 0.2$ (*1*), 0.5 (*2*) and 0.75 (*3*), $nE_{sw} = 50\,\mathrm{mV}$ and $n\Delta E = -2\,\mathrm{mV}$

$\kappa > 10$, quasireversible if $0.01 < \kappa < 10$ and irreversible if $\kappa < 0.01$. In the quasireversible range $\Delta\Psi_p$ decreases with decreasing κ. This is partly caused by the transformation of the backward component under the influence of increased frequency (see Fig. 2.5). The maximum of this component decreases faster than the absolute value of the minimum of the forward component. For the conditions of Fig. 2.5, the ratio $I_{p,f}/I_{p,b}$ decreases from -1.47, for the reversible reaction (see Table 2.1), to -2.25 for $\kappa = 0.3$, and to -15.4 for $\kappa = 0.05$. The difference of the peak potentials of the components, $E_{p,b} - E_{p,f}$, changes from $10\,\mathrm{mV}$, for $\kappa = 0.3$, to $110\,\mathrm{mV}$ for $\kappa = 0.05$. If the reaction is totally irreversible, the backward component is negative for all potentials and its maximum disappears (see Fig. 2.6). So, the net response of irreversible reaction is smaller than the absolute value of its forward component [16, 18].

The half-peak width, $\Delta E_{p/2}$, of the quasireversible reaction increases with decreasing κ, but $\Delta E_{p/2}$ of totally irreversible reactions is independent of κ.

The net peak current, ΔI_p, of the quasireversible reaction is not a linear function of the square-root of frequency. This is shown in Fig. 2.8. In principle, the gradient $\partial(\Delta I_p/nFAc_O^*D^{1/2})/\partial f^{1/2}$ may change from $(\Delta\Psi_p)_{\mathrm{revers.}}$, at the lowest frequency, to $(\Delta\Psi_p)_{\mathrm{irrevers.}}$ at the highest frequency (see straight lines a and b, respectively, in Fig. 2.8). However, the frequency can be varied only within rather narrow range, from 10–2000 Hz, so that the parameter κ can not be changed more than fourteen times. For this reason the dependence of ΔI_p on $f^{1/2}$ for the particular electrode reaction is a certain fraction of the general relationship. There are two limiting cases: the reaction is apparently reversible at lower frequencies and quasireversible at higher frequencies (see curve 1 in Fig. 2.8), or the reaction appears quasireversible at lower frequencies and totally irreversible at higher frequencies (see curve 3 in Fig. 2.8).

The dimensionless net peak current of totally irreversible electrode reaction is a linear function of the transfer coefficient: $\Delta\Psi_p = 0.235\alpha$. This relationship is shown in Fig. 2.9.

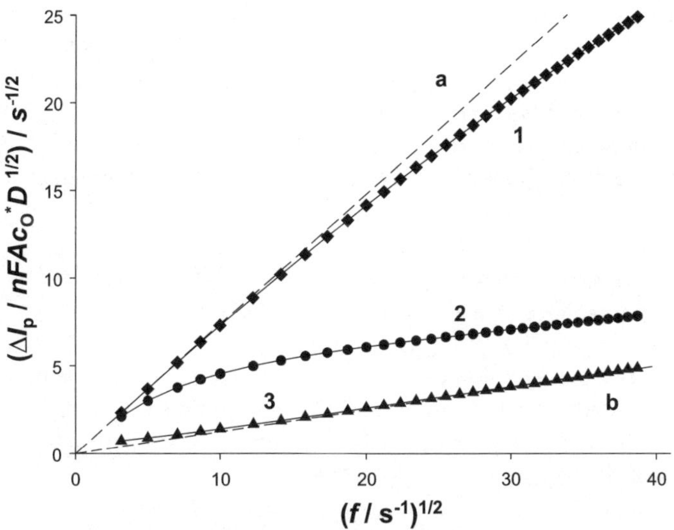

Fig. 2.8 The dependence of normalized net peak current on the square-root of frequency; $k_s = 0.1$ (*1*), 0.01 (*2*) and 0.001 cm/s (*3*), $\alpha = 0.5$, $D = 9 \times 10^{-6}\,\mathrm{cm^2/s}$, $nE_{sw} = 50\,\mathrm{mV}$ and $n\Delta E = -2\,\mathrm{mV}$. The *broken lines* correspond to reversible (*a*) and totally irreversible reactions (*b*)

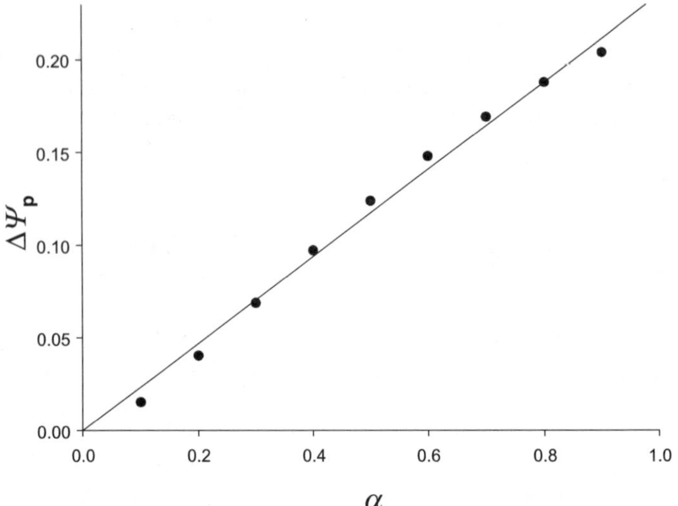

Fig. 2.9 The dependence of the dimensionless net peak current of irreversible electrode reaction (1.1) on the transfer coefficient; $\kappa = 10^{-3}$, $nE_{sw} = 50\,\mathrm{mV}$ and $n\Delta E = -2\,\mathrm{mV}$

Figure 2.10 shows the dependence of peak potentials on the logarithm of the parameter κ. In the upper quasireversible range ($0.5 < \kappa < 10$) an interesting inversion of peak potentials of the components can be noted (see Fig. 2.10a). In this range the response is very sensitive to a change in the signal parameters [19–21]. If $\kappa < 0.04$, the maximum of the backward component is too small to be noticed experimentally. The net peak potential E_p and the peak potential of the forward component $E_{p,f}$ of irreversible reactions ($\kappa < 0.01$) are both linear functions of the logarithm of the parameter κ with the slope $2.3RT/\alpha nF$ (see Fig. 2.10b). Considering the definition of κ, this means that these peak potentials depend linearly on the logarithm of frequency, with the slope $-2.3RT/2\alpha nF$. If $nE_{sw} = 50\,\mathrm{mV}$ and $n\Delta E = -2\,\mathrm{mV}$, the relationship between E_p and $\log \kappa$ is defined by the following equation:

$$E_p - E^{\theta} = 2.3(RT/\alpha nF)\log \kappa + 0.044/\alpha n \tag{2.3}$$

Equation (2.3) was calculated from the linear relationship between the net peak potential and the reciprocal of the product αn, which is shown in Fig. 2.11. It was calculated for $\kappa = 0.001$, $0.1 \leq \alpha \leq 0.9$ and $n = 1$, 2 and 3. The slope of this straight line is $2.3(RT/F)\log \kappa + 0.044$. The intersection of (2.3) with the straight line $E_p = E^{\theta}$ corresponds to the critical kinetic parameter $\log \kappa_0 = -0.75$ (see Fig. 2.10b). If the standard potential E^{θ} is known, the standard rate constant can be determined by the variation of frequency:

$$\log k_s = -0.75 + \log D^{1/2} + \log f_0^{1/2} \tag{2.4}$$

where f_0 is the frequency of intersection. However, the range of frequencies is limited and all parameters of a certain electrode reaction (E^{θ}, n, α, D and k_s) can not be determined by the variation of frequency.

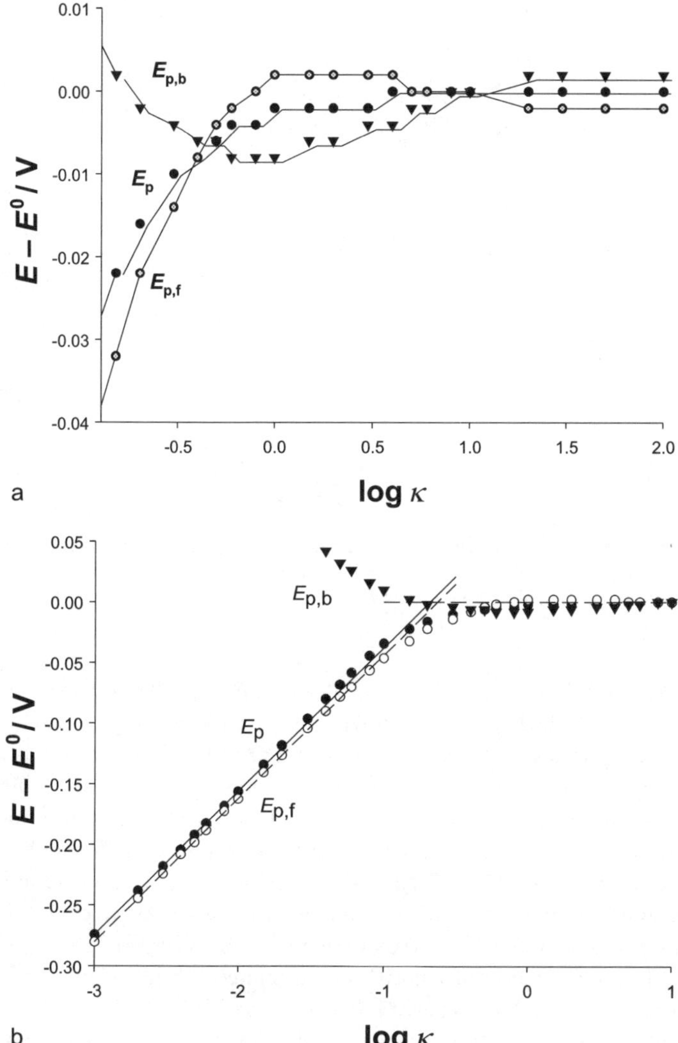

Fig. 2.10a,b The dependence of peak potentials on the logarithm of the dimensionless kinetic parameter; $\alpha = 0.5$, $nE_{\mathrm{sw}} = 50\,\mathrm{mV}$ and $n\Delta E = -2\,\mathrm{mV}$. A reversible to upper quasireversible range (**a**) and the quasireversible to irreversible range (**b**)

If $\alpha < 0.3$, a split net response of quasireversible reaction may appear within a certain range of frequencies [15, 17]. As can be seen in Fig. 2.12, the backward component is characterized by a small maximum and a deep and broad minimum. The net response is determined by the maximum of the backward component and by the difference in minima of the backward and forward components. So, it may consist of two peaks which are marked as I and II in Fig. 2.12. The peak II appears for $\kappa < 0.06$, but its maximum current is smaller than the maximum current of the peak I if $\kappa \geq 0.015$. The peak I disappears for $\kappa < 10^{-3}$. Formally, the net peak

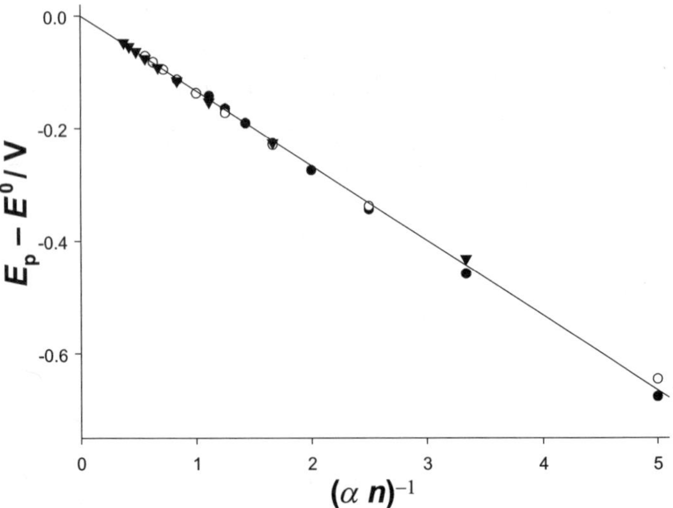

Fig. 2.11 The dependence of the net peak potential of irreversible reaction on the reciprocal of the product of the transfer coefficient and the number of electrons; $\kappa = 10^{-3}$, $nE_{sw} = 50$ mV, $n\Delta E = -2$ mV and $n = 1$ ($\bullet$), 2 ($\circ$) and 3 ($\blacktriangledown$)

potential E_p abruptly decreases from 0.070 V vs. E^{θ} to -0.696 V within a narrow range $0.015 > \kappa > 0.01$, which is shown in Fig. 2.13. However, this is the consequence of the split net response and the potential of maximum of the peak II (the crosses in Fig. 2.13) follows the potential of minimum of the forward component. In the lower quasireversible range ($0.01 < \kappa < 0.1$) the peak potential of the forward component should be used for the determination of transfer coefficient by the variation of frequency (see also Fig. 2.10b). If $\alpha = 0.2$, the net peak potential is close to the peak potential of the backward component for $\kappa > 0.04$, but if $\kappa < 0.04$ it is close to the peak potential of the forward component. An experimental example is shown in Fig. 2.14. The reduction of zinc(II) ions at mercury electrode was measured at various square-wave frequencies [21]. If the frequency is 200 Hz, or lower, the net peak potential is about -1 V, but at 500 Hz it is about -1.11 V. The peaks marked as I and II in Fig. 2.12 can be identified. Kinetic parameters of this reaction were determined by a non-linear least squares method: $k_s = (2.64 \pm 0.16) \times 10^{-4}$ cm/s, $\alpha = 0.20 \pm 0.02$ and $E_{1/2,r} = 1.000 \pm 0.001$ V vs. SCE. Figure 2.15 shows that theoretical currents calculated with these parameters fit well to the forward and backward components of the experimental voltammogram.

The split net response may also appear if square-wave voltammogram of irreversible electrode reaction (1.1) is recorded starting from low potential, at which the reduction is diffusion controlled [22, 23]. This is shown in Fig. 2.16b. If the starting potential is 0.3 V vs. E^{θ}, a single net peak appears and the backward component of the response does not indicate the re-oxidation of the product (see Fig. 2.16a). If the reverse scan is applied ($E_{st} = -0.8$ V, Fig. 2.16b), the forward, mainly oxidative component Ψ_f is in maximum at 0.190 V, while the backward, partly reductive

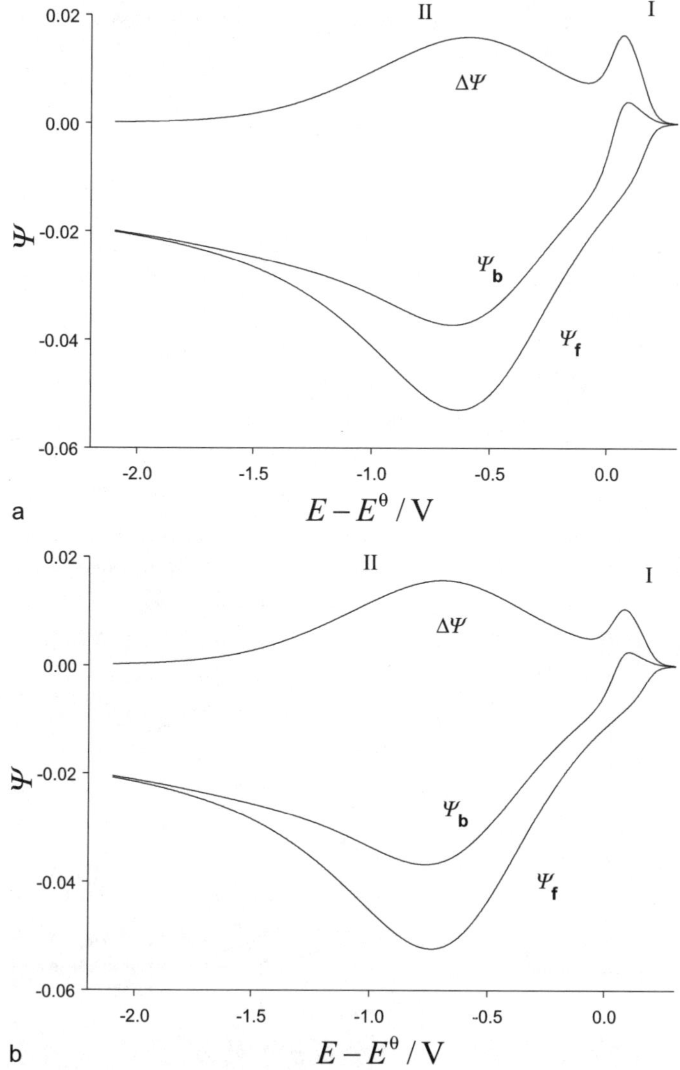

Fig. 2.12a,b SWV of quasireversible electrode reaction (1.1); $\kappa = 0.015$ (**a**) and 0.01 (**b**), $\alpha = 0.1$, $nE_{sw} = 50$ mV, $n\Delta E = -2$ mV and $E_{st} - E^{\theta} = 0.3$ V

component Ψ_b is in minimum at -0.292 V. Because of this separation of the extremes of components, the net response consists of two peaks: the reductive one at lower potential and the oxidative one at higher potial. The split net response appears because a certain amount of product is created at the starting potential. The rate of reduction of the reactant is lower at the anodic, forward series of pulses (see the forward component for $E < -0.1$ V) and higher at the cathodic, backward series of pulses. This is the reason for the minimum of the backward component. At potentials higher than E^{θ}, the remaining product is oxidized and the maximum

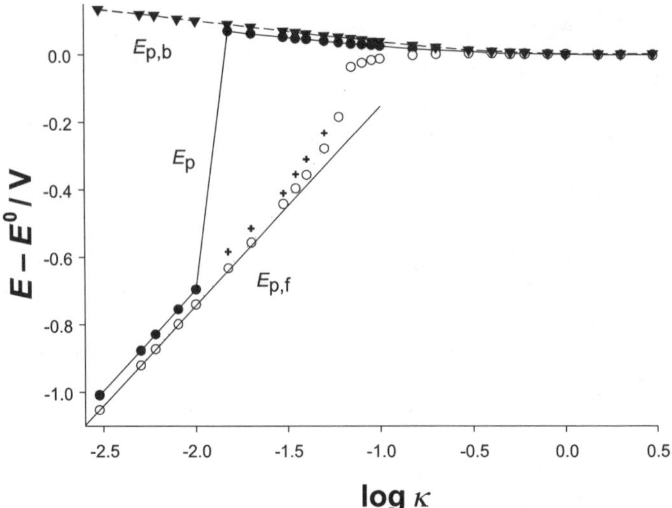

Fig. 2.13 The dependence of peak potentials on the logarithm of dimensionless kinetic parameter; $\alpha = 0.1$, $nE_{sw} = 50$ mV and $n\Delta E = -2$ mV. The *crosses* denote the potentials of maxima of the peak II

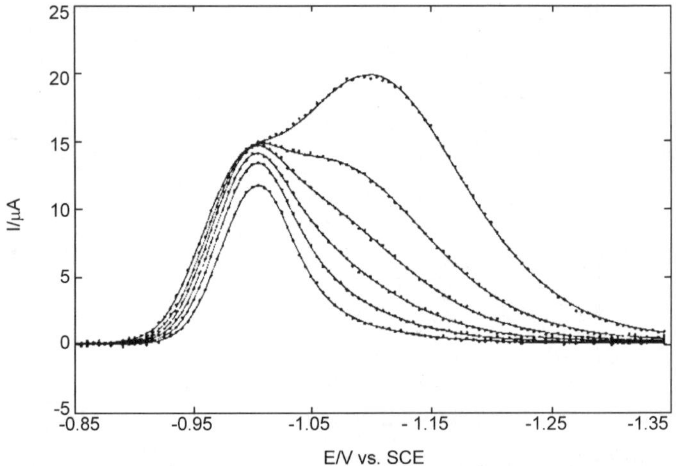

Fig. 2.14 Square-wave voltammograms for reduction of 1 mM Zn(II) in 1.0 M KNO$_3$. $\Delta E = 5$ mV, $E_{sw} = 25$ mV. Experimental (......) and best fitting theoretical (———) currents with f in ascending order of curves: 10, 25, 50, 100, 200, 500 Hz (reprinted from [21] with permission)

of the forward component appears. The reverse scan can be used to analyze the mechanism of electrode reaction [24, 25]. Figure 2.17 shows square-wave voltammograms of 5×10^{-4} M europium(III) at a static mercury drop electrode in 0.1 M NaClO$_4$ acidified by 10^{-2} M HClO$_4$ [23]. If the scan direction is negative, the reduction Eu^{3+} + e$^-$ $\rightarrow$ Eu^{2+} appears totally irreversible, but the reverse scan reveals that Eu^{2+} is oxidized at about -0.35 V vs. Ag/AgCl (sat. NaCl).

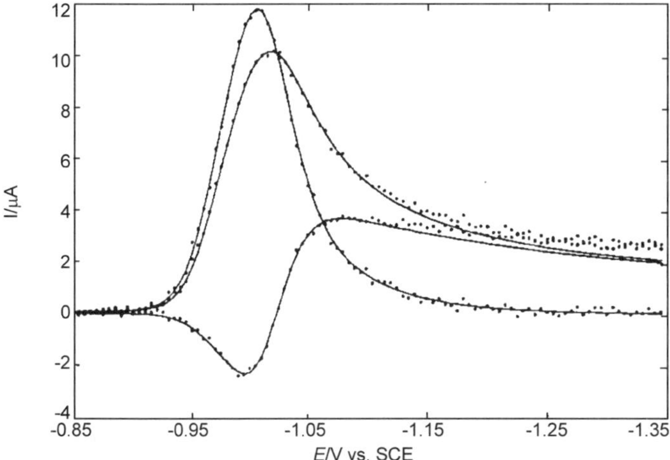

Fig. 2.15 Square-wave voltammogram collected at 10 Hz, background voltammogram subtracted. 1 mM Zn(II) in 1.0 M KNO$_3$. $\Delta E = 5$ mV, $E_{sw} = 25$ mV. Experimental (......) and best fitting theoretical (——) currents. Theoretical forward and reverse currents calculated from the best fit to the experimental net currents using the semi-infinite planar diffusion equation and boundary conditions (reprinted from [21] with permission)

2.2 Reactions of Dissolved Species on Spherical Electrodes and Microelectrodes

On a stationary spherical electrode, the reaction (1.1) can be mathematically represented by the following system of differential equations and boundary conditions:

$$\partial(rc_O)/\partial t = D\,\partial^2(rc_O)/\partial r^2 \tag{2.5}$$

$$\partial(rc_R)/\partial t = D\,\partial^2(rc_R)/\partial r^2 \tag{2.6}$$

$$t = 0, \quad r \geq r_0: \quad c_O = c_O^*, \quad c_R = 0 \tag{2.7}$$

$$t > 0, \quad r \to \infty: \quad c_O \to c_O^*, \quad c_R \to 0 \tag{2.8}$$

$$r = r_0: \quad D(\partial c_O/\partial r)_{r=r_0} = -I/nFA \tag{2.9}$$

$$D(\partial c_R/\partial r)_{r=r_0} = I/nFA \tag{2.10}$$

where r_0 is the electrode radius and r is the distance from the centre of electrode. The meaning of other symbols is given below (1.9). Equation (2.5) is a condensed form of the equation:

$$\frac{\partial c_O}{\partial t} = D\left(\frac{\partial^2 c_O}{\partial r^2} + \frac{2}{r}\cdot\frac{\partial c_O}{\partial r}\right) \tag{2.11}$$

which was obtained by the transformation:

$$\frac{\partial^2(rc_O)}{\partial r^2} = r\frac{\partial^2 c_O}{\partial r^2} + 2\frac{\partial c_O}{\partial r} \tag{2.12}$$

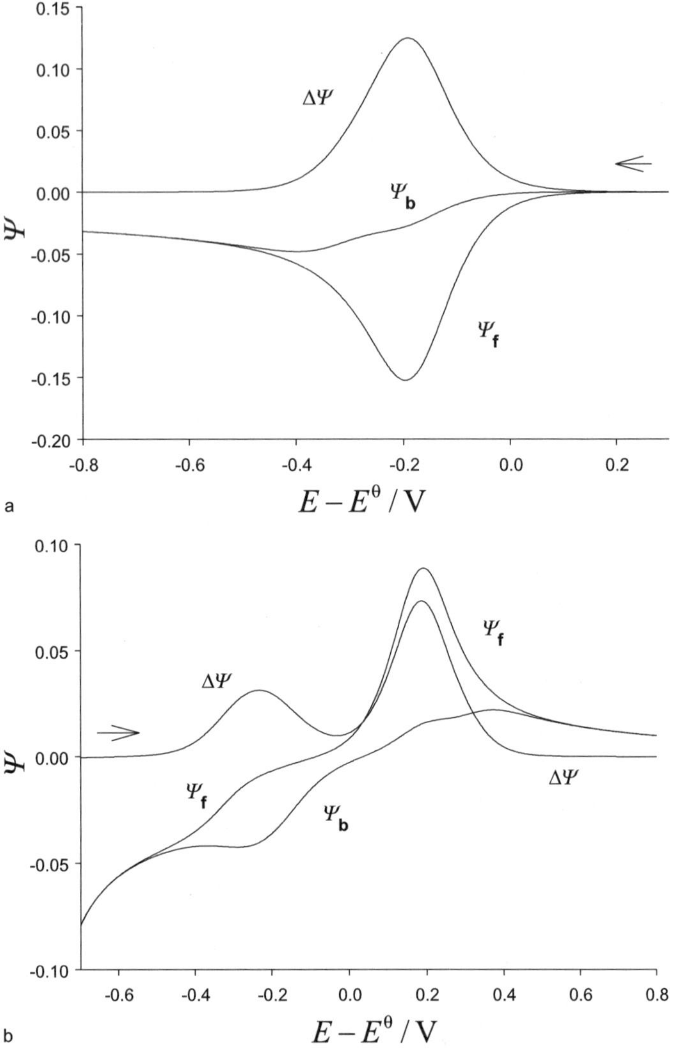

Fig. 2.16a,b SWV of irreversible electrode reaction (1.1); $E_{st} - E^\theta = 0.3$ (**a**) and -0.8 V (**b**), $\kappa = 0.005$, $\alpha = 0.5$, $nE_{sw} = 50$ mV and $n\Delta E = -2$ (**a**) and 2 mV (**b**)

The solution of (2.5) and (2.6) is a system of integral equations:

$$(c_O)_{r=r_0} = c_O^* + J \tag{2.13}$$

$$(c_R)_{r=r_0} = -J \tag{2.14}$$

$$J = \int_0^t \frac{I}{nFA\sqrt{D}} \left[\frac{1}{\sqrt{\pi(t-\tau)}} - \frac{\sqrt{D}}{r_0} \exp\cdot\frac{D(t-\tau)}{r_0^2} \cdot \mathrm{erfc}\frac{\sqrt{D(t-\tau)}}{r_0} \right] d\tau \tag{2.15}$$

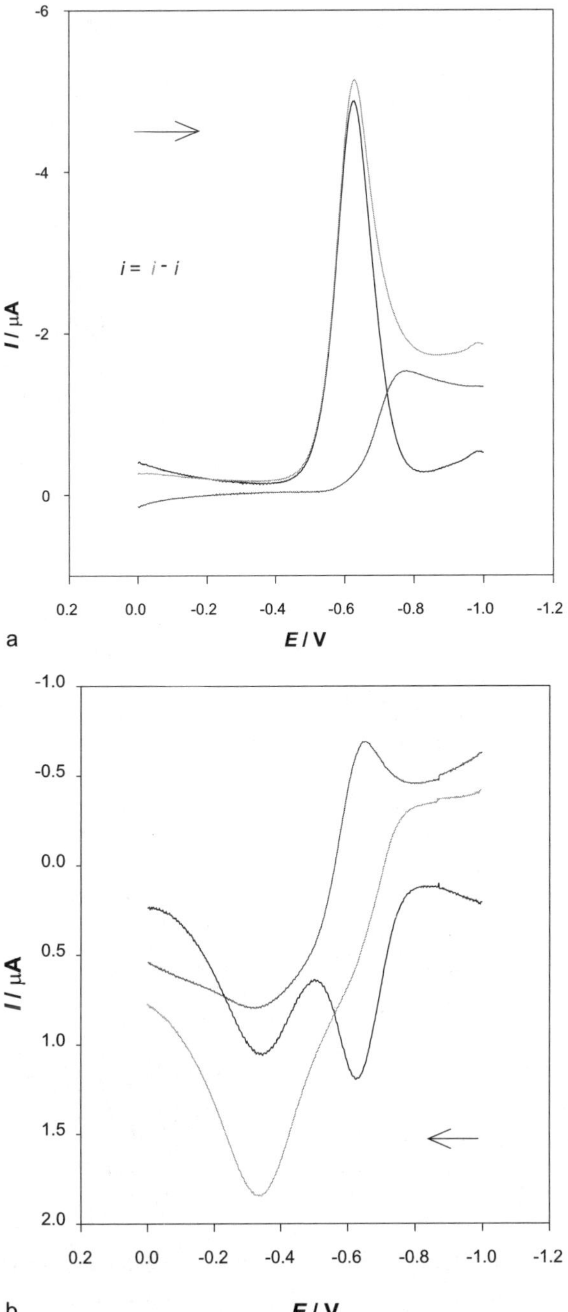

Fig. 2.17a,b Square-wave voltammogram of Eu^{3+} (0.5 mmol dm^{-3}) in acidified 0.1 mol dm^{-3} NaClO$_4$ and its forward (*red*) and backward (*blue*) currents. Frequency: 125 s^{-1}; amplitude: 40 mV; step potential: 2 mV; delay time: 30 s; scan direction: negative (**a**) and positive (**b**) (reprinted from [23] with permission)

For the numerical integration the following transformation is used:

$$J = \frac{1}{\rho} \sum_{j=1}^{m} \left(\frac{I}{nFA\sqrt{Df}} \right)_j S_{m-j+1} \tag{2.16}$$

$$S_1 = 1 - \exp\frac{\rho^2}{50} \cdot \text{erfc}\frac{\rho}{5\sqrt{2}} \tag{2.17}$$

$$S_k = \exp\frac{\rho^2(k-1)}{50} \cdot \text{erfc}\frac{\rho\sqrt{k-1}}{5\sqrt{2}} - \exp\frac{\rho^2 k}{50} \cdot \text{erfc}\frac{\rho\sqrt{k}}{5\sqrt{2}} \tag{2.18}$$

where $\rho = \sqrt{D}/r_0\sqrt{f}$. If reaction (1.1) is electrochemically reversible, then (2.13) and (2.14) are introduced into the Nernst equation (1.8) and the following system of recursive formulae is obtained [26]:

$$\Psi_1 = -\frac{\rho}{S_1\left[1 + \exp(\varphi_1)\right]} \tag{2.19}$$

$$\Psi_m = -\frac{\rho + \left[1 + \exp(\varphi_m)\right]\sum\limits_{j=1}^{m-1}\Psi_j S_{m-j+1}}{S_1\left[1 + \exp(\varphi_m)\right]} \tag{2.20}$$

where $\Psi = I/nFAc_O^*(Df)^{1/2}$ is dimensionless current and the meaning of other symbols is given below (1.24).

The application of (2.17)–(2.20) is shown in Fig. 2.18. The response depends on the sphericity parameter $\rho = \sqrt{D}/r_0\sqrt{f}$ [27]. Under the influence of increasing parameter ρ, the minimum of the forward component and the maximum of the backward component gradually vanish and both components acquire the form of a polarographic wave. At potentials much lower than the half-wave potential, both currents tend to the limiting value which is equal to $-\rho$. The net peak potential is equal to the reversible half-wave potential and independent of the sphericity parameter, but the dimensionless net peak current is a linear function of the parameter ρ. If $nE_{sw} = 50\,\text{mV}$ and $n\Delta E = -5\,\text{mV}$, this relationship is:

$$\Delta\Psi_p = 0.752 + 0.749\rho \tag{2.21}$$

This is shown in Fig. 2.19. The relationships between $\Delta\Psi_p$ and nE_{sw} and $n\Delta E$ do not depend on electrode size [28–30]. So, if $nE_{sw} = 25\,\text{mV}$ and $n\Delta E = -5\,\text{mV}$, the relationship (2.21) is: $\Delta\Psi_p = 0.465 + 0.45\rho$ [26]. If the frequency is high and a hanging mercury drop electrode is used, the spherical effect is usually negligible ($\rho < 10^{-2}$). However, the influence of sphericity must be taken into consideration under most other conditions, and generally at microelectrodes. The net peak current is a linear function of the square-root of frequency: $\Delta I_p/nFAc_O^*D^{1/2} = 0.752 \times f^{1/2} + 0.749 \times D^{1/2}/r_0^{1/2}$. This relationship is shown in Fig. 2.20, for $D = 9 \times 10^{-6}\,\text{cm}^2/\text{s}$ and $r_0 = 10^{-3}\,\text{cm}$. Considering that the surface area of a hemispherical electrode is $A = 2\pi r_0^2$, the net peak current can be expressed as:

$$\Delta I_p = 4.706\,nFc_O^* D^{1/2} r_0 [1.004\,r_0 f^{1/2} + D^{1/2}] \tag{2.22}$$

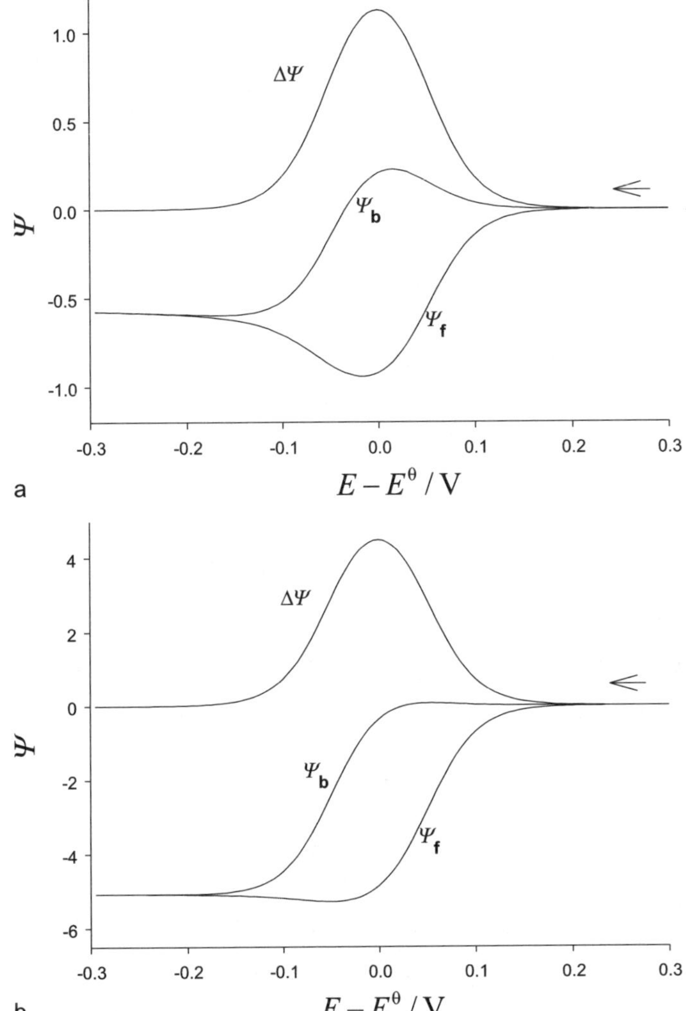

Fig. 2.18a,b SWV of reversible reaction (1.1) on spherical electrode; $\rho = 0.5$ (**a**) and 5 (**b**), $nE_{\mathrm{sw}} =$ 50 mV and $n\Delta E = -5$ mV

Theoretically, if an extremely small electrode is used and low frequency is applied, so that $r_0 f^{1/2} \ll D^{1/2}$, a steady-state, frequency-independent net peak current should appear: $\Delta I_{\mathrm{p,ss}} = 4.706 nF c_{\mathrm{O}}^{*} D r_0$.

At an inlaid microdisk electrode, the dependence of the dimensionless net peak current on the sphericity parameter is given by the following equation [31]:

$$\Delta \Psi_{\mathrm{p}} = 0.598 + 0.141 \exp(-1.6\rho) + 0.955\rho \tag{2.23}$$

However, at moderately small electrodes a linear relationship exists:

$$\Delta \Psi_{\mathrm{p}} = 0.47 + 0.77\rho \tag{2.24}$$

These relationships were calculated for $nE_{\mathrm{sw}} = 50$ mV and $n\Delta E = -10$ mV [31].

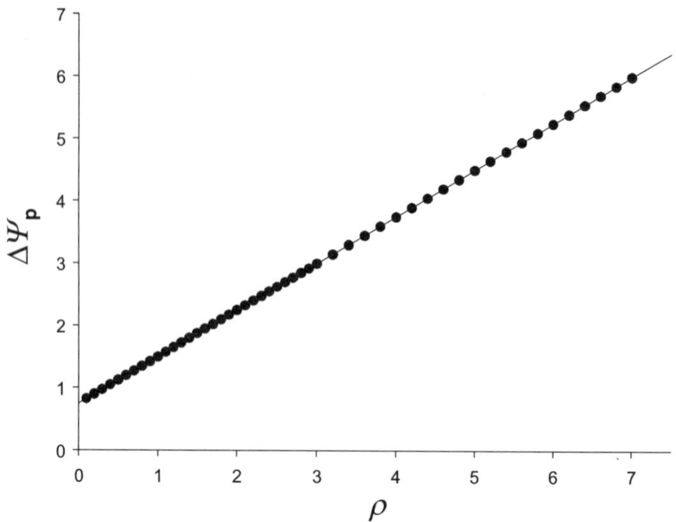

Fig. 2.19 Dependence of dimensionless net peak current of reversible reaction (1.1) on the sphericity parameter; $nE_{sw} = 50\,mV$ and $n\Delta E = -5\,mV$

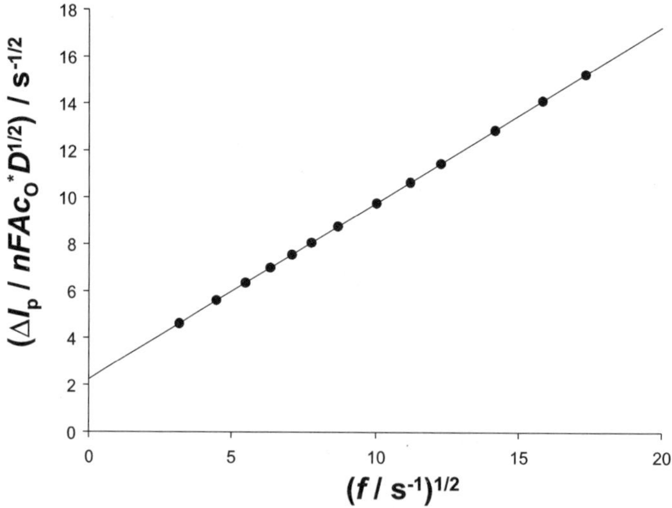

Fig. 2.20 Dependence of normalized net peak current of reversible reaction (1.1) on the square-root of frequency; $nE_{sw} = 50\,mV$, $n\Delta E = -5\,mV$, $D = 9 \times 10^{-6}\,cm^2/s$ and $r_0 = 10^{-3}\,cm$

If the electrode reaction (1.1) is kinetically controlled, the response depends on both the parameter ρ and the kinetic parameter κ [26, 27]. If the electrode size is constant and the frequency is varied, both parameters ρ and κ are changed. Also, if a certain reaction is measured at constant frequency, with a range of microelectrodes having various diameters, the apparent reversibility of the reaction decreases with the decreasing diameter because of radial diffusion. So, the relationship between the dimensionless net peak current of quasireversible electrode reactions and the sphericity parameter ρ is not linear [26]. If the reaction is totally irreversible, $\Delta\Psi_p$ is independent of κ, but its relationship with ρ is a curve characterized by two asymptotes: $\Delta\Psi_p = 0.11 + 0.32\rho$ (for $\rho < 0.5$) and $\Delta\Psi_p = 0.15 + 0.24\rho$ (for $0.5 < \rho < 10$), both for $nE_{sw} = 25$ mV, $n\Delta E = -5$ mV and $\alpha = 0.5$ [26]. The slopes and intercepts of these straight lines are linear functions of the transfer coefficient α. Thus, the general relationship between $\Delta\Psi_p$ and αn, described for planar electrodes (see Fig. 2.9), holds for spherical electrodes as well.

Figure 2.21 shows the dependence of dimensionless net peak currents of ferrocene and ferricyanide on the sphericity parameter (note that $\Delta\Phi_p = \Delta\Psi_p$ and $y = \rho$). The SWV experiments were performed at three different gold inlaid disk electrodes ($r_0 = 30$, 12.5 and 5 μm) and the frequencies were changed over the range from 20 to 2000 Hz [26]. For ferrocene the relationship between $\Delta\Psi_p$ and ρ is linear: $\Delta\Psi_p = 0.88 + 0.74\rho$. This indicates that the electrode reaction of ferrocene is electrochemically reversible regardless of the frequency and the electrode radius over the range examined. For ferricyanide the dependence of $\Delta\Psi_p$ on ρ appears in sequences. Each sequence corresponds to a particular value of the parameter $D^{1/2}/r_0$. The results obtained with the same frequency, but at different microelectrodes, are connected with thin, broken lines. The difference in the responses of these

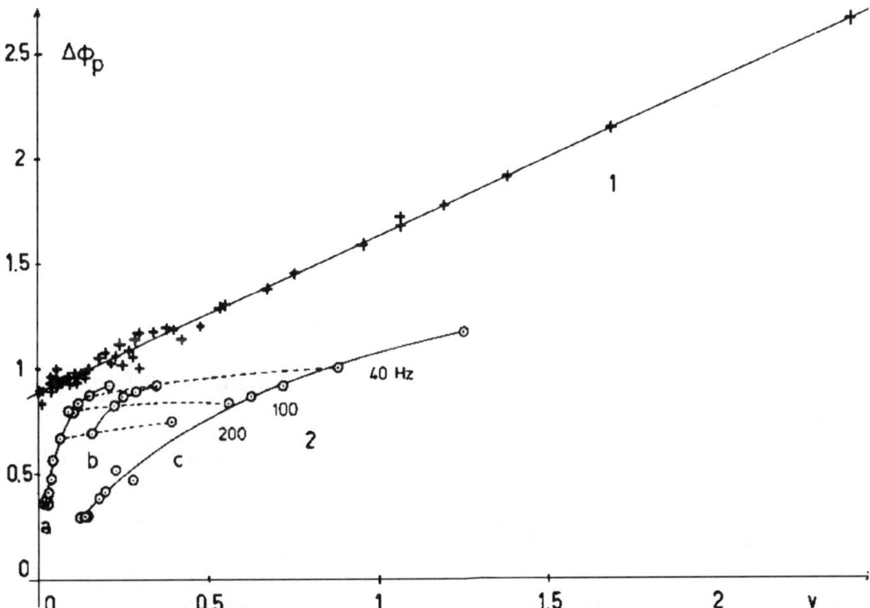

Fig. 2.21 Dimensionless SW peak currents for oxidation of ferrocene (*1*) and reduction of ferricyanide (*2, a–c*) as functions of the parameter $y = (D/f)^{1/2}/r_0$. $\Delta\Phi_p = \Delta I_p(FSc^*)^{-1}(Df)^{-1/2}$; $c^*/\text{mol cm}^{-3} = 10^{-7}$ (*1*) and 5×10^{-6} (*2*); $D/\text{cm}^2 \text{ s}^{-1} = 2.3 \times 10^{-5}$ (*1*) and 7.8×10^{-6} (*2*); $\Delta E = 2$ mV; $E_{sw} = 50$ mV; $20 < f/\text{Hz} < 2000$; and $r_0/\text{cm} = 3 \times 10^{-3}$ (*2a*), 1.25×10^{-3} (*2b*) and 5×10^{-4} (*2c*) (reprinted from [26] with permission)

two electrode reactions was ascribed to the differences in the standard rate constants and the diffusion coefficients of ferrocene in acetonitrile ($k_s \geq 6\,\mathrm{cm/s}$ [32], $D = 2.3 \times 10^{-5}\,\mathrm{cm^2/s}$ [33]) and ferricyanide in aqueous electrolyte ($k_s = 0.02\,\mathrm{cm/s}$, $D = 7.8 \times 10^{-6}\,\mathrm{cm^2/s}$ [34]). So, the SWV responses of ferricyanide at microelectrodes appear quasi-reversible [26].

Electroanalytical application of hemispherical [35, 36], cylindrical [37, 38] and ring microelectrodes [39] has been described. A hemispherical iridium-based mercury ultramicroelectrode was formed by coulometric deposition at $-0.2\,\mathrm{V}$ vs. SSCE in solution containing $8 \times 10^{-3}\,\mathrm{M}$ Hg(II) and $0.1\,\mathrm{M}$ HClO$_4$ [35]. The radius of the iridium wire was $6.5\,\mu\mathrm{m}$. The electrode was used for anodic stripping SWV determination of cadmium, lead and copper in unmodified drinking water, without any added electrolyte, deoxygenation, or forced convection. The effects of finite volume and sphericity of mercury drop electrode in square-wave voltammetry have been also studied [36].

Microcylindrical electrodes are easier to construct and maintain than microdisk electrodes [37]. Mass transport to a stationary cylinder in quiescent solution is governed by axisymmetrical cylindrical diffusion. For square-wave voltammetry the shape and position of the net current response are independent of the extent of cylindrical diffusion [38]. The experiments were performed with the ferri-ferrocyanide couple using a small platinum wire ($25\,\mu\mathrm{m}$ in diameter and $0.5-1.0\,\mathrm{cm}$ in length) as the working electrode [37].

Ring microelectrodes have been shown to exhibit high current density and high signal-to-background and signal-to-noise ratios, which is important in analytical voltammetry. In square-wave voltammetry electrode geometry has negligible influence on peak position and peak width, but has a significant influence on peak height [39].

2.3 Reactions of Amalgam-Forming Metals on Thin Mercury Film Electrodes

2.3.1 Reversible Reduction of Metal Ions on Stationary Electrode

Figure 2.22 shows SWV responses of electrochemically reversible reaction on stationary planar electrodes covered with a thin mercury film:

$$\mathrm{M^{n+}} + n\,\mathrm{e^-} \leftrightarrows \mathrm{M^o(Hg)} \tag{2.25}$$

The voltammograms were calculated by assuming that no metal atoms were initially present in the film. The response depends on the dimensionless film thickness $\lambda = L(f/D)^{1/2}$, where L is the real film thickness [40]. Figure 2.23a shows that the dimensionless net response is the highest if $\lambda = 1$. This condition is satisfied if, for instance, $D = 9 \times 10^{-6}\,\mathrm{cm^2/s}$, $f = 100\,\mathrm{Hz}$ and $L = 3\,\mu\mathrm{m}$, which is rather thick film. In trace analysis the films are usually much thinner than a micrometer [41]. The maximum net response appears if the film thickness is approximately equal to

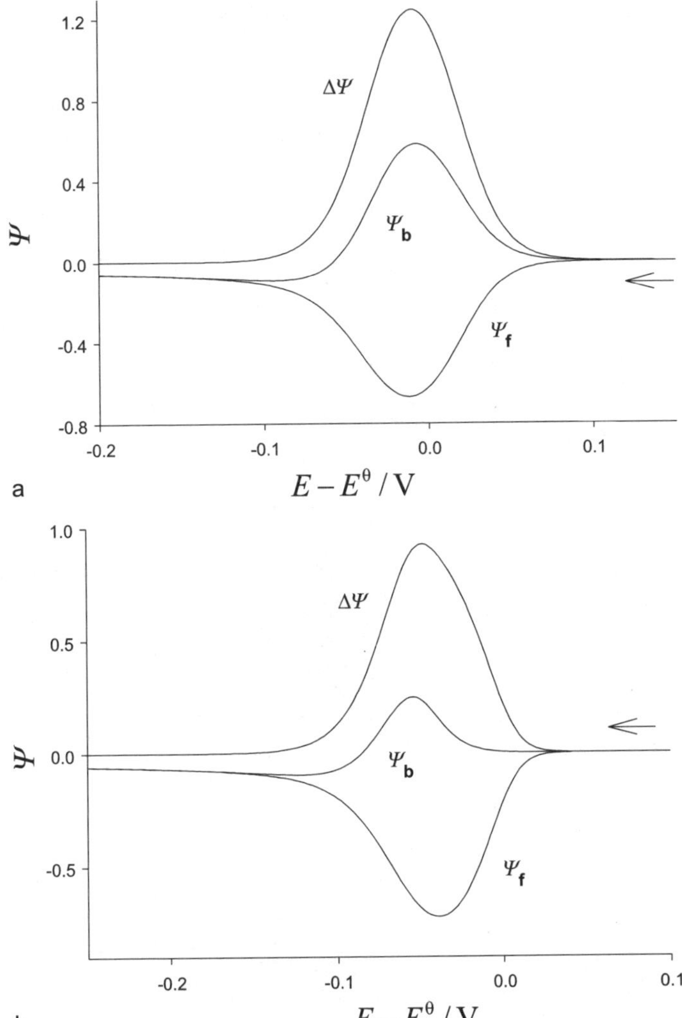

Fig. 2.22a,b SWV of the reversible amalgam-forming reaction (2.25) on the thin mercury film electrode; $\lambda = 1.118$ (**a**) and 0.112 (**b**), $n\Delta E = -4\,\mathrm{mV}$ and $nE_{\mathrm{sw}} = 50\,\mathrm{mV}$

the diffusion layer thickness of metal atoms in the film [42]. Figure 2.23a shows that the maximum is more pronounced if the square-wave amplitude is higher. For the parameters of Fig. 2.22, the ratio of extremes of the components changes from $I_{\mathrm{p,f}}/I_{\mathrm{p,b}} = -1.7$, for $\lambda > 10$, to -1.15, for $\lambda = 1$, and -2.9 for $\lambda < 0.3$. So, the maximum is caused mainly by the increase and decrease of the peak current of the backward component.

The parameter λ is a linear function of the square-root of frequency but $\Delta\Psi_{\mathrm{p}}$ depends on λ only within the interval $0.1 < \lambda < 10$. So, the net peak current ΔI_{p} depends linearly on $f^{1/2}$ if the change of frequency does not cause the change of

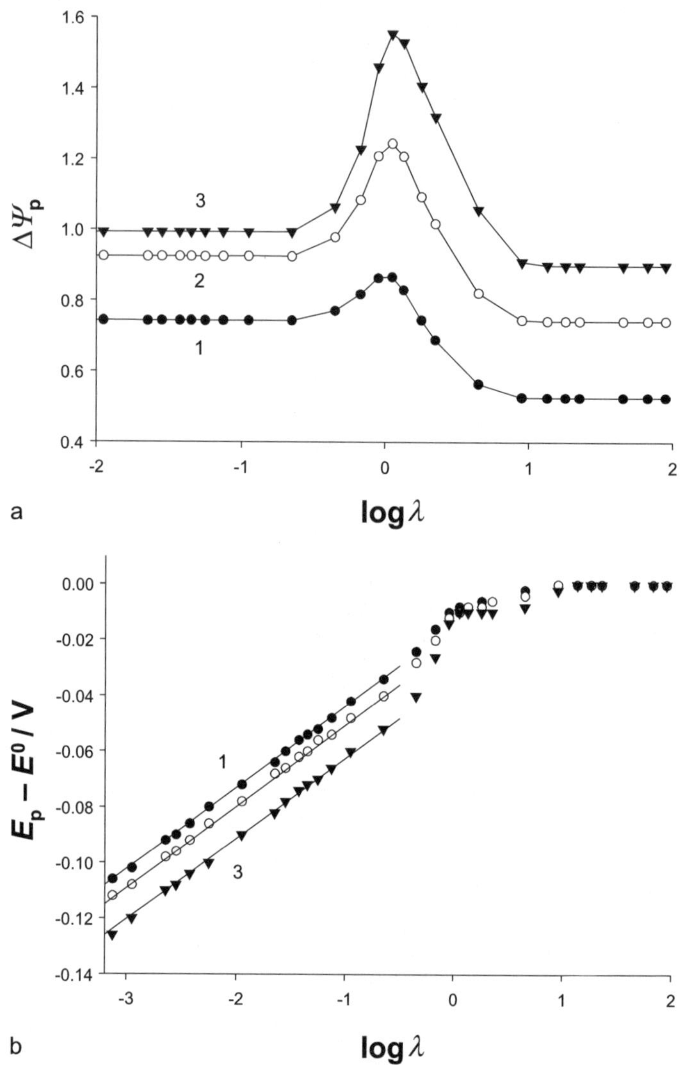

Fig. 2.23a,b Dependence of dimensionless net peak currents (**a**) and the net peak potentials (**b**) on the logarithm of dimensionless film thickness; $n\Delta E = -4\,\mathrm{mV}$ and $nE_{\mathrm{sw}}/\mathrm{mV} = 30$ (*1*), 50 (*2*) and 80 (*3*)

$\Delta\Psi_{\mathrm{p}}$. This means that the linear relationship between ΔI_{p} and $f^{1/2}$ exists if either $\lambda < 0.1$, or $\lambda > 10$ at all frequencies [40, 42].

Figure 2.23b shows the dependence of the net peak potentials on the logarithm of the parameter λ. If the film is very thin ($\lambda < 0.1$), this relationship is linear, with the slope $\partial E_{\mathrm{p}}/\partial \log \lambda = 2.3RT/nF$. On very thick films ($\lambda > 10$) the net peak potential is independent of λ, regardless of square-wave amplitude. So, if at the lowest frequency $\lambda > 10$, the net peak potential is independent of frequency. On the contrary, if $\lambda < 0.1$ at the highest frequency, E_p is a linear function of the logarithm of frequency, with the slope $\partial E_p/\partial \log f = 2.3RT/2nF$. If these conditions are not satisfied, the gradient of the relationship between E_p and $\log f$ changes from $2.3RT/2nF$ to zero in a complex manner, as can be seen in Fig. 2.23b [40].

2.3.2 Anodic Stripping Square-Wave Voltammetry of Metal Ions

Anodic stripping voltammetry is commonly applied to the analytical determination of a wide range of trace metals capable of forming an amalgam. The method has two stages: first, a preconcentration step is performed in which electrodeposition of metal ions in solution leads to the accumulation of metal as an amalgam. Second, the electrode potential is swept to positive potentials, inducing the oxidation of the metal in the mercury electrode. The highest sensitivity is obtained if a thin mercury film covered rotating disk electrode is used in the combination with SWV as a stripping technique. On this electrode the accumulation is performed under hydrodynamic conditions, which provide effective and stable mass transfer during this step [43], but usually the rotating of the electrode must be stopped before the stripping peaks are recorded in order to decrease the electrical noise. So, a short rest period is introduced between two steps to allow the solution to calm down. The factor of preconcentration is inversely proportional to the film thickness:

$$c_{M(Hg)}/c^* = Dt_{acc}/L\delta \tag{2.26}$$

where $c_{M(Hg)}$ and c^* are the concentrations of metal atoms in mercury and metal ions in the bulk of the solution, respectively, t_{acc} is a duration of accumulation and δ is the thickness of the diffusion layer at the rotating electrode during the accumulation period.

Simulations of anodic stripping SWV responses of the reaction (2.25) are shown in Fig. 2.24 (Ψ is defined as below (1.24)), and the dependence of dimensionless net peak current on the logarithm of the parameter λ is shown in Fig. 2.25a. In the calculation it was assumed that the accumulation time was 60 s, the rest period was 10 s, the potential of the accumulation and the starting potential of the stripping scan were both -0.3 V vs. E^θ and the diffusion layer thickness during the accumulation was 10^{-3} cm. If the mercury film is rather thick ($\lambda > 1$), $\Delta\Psi_p$ decreases with increasing value of λ in agreement with (2.26). However, under the assumed conditions, the maximum factor of preconcentration is 3×10^4, appearing for $\lambda = 1$, while on thin films ($\lambda < 0.3$) the factor is 2.5×10^4. Phenomenologically, this can be explained by the change of the form of the forward (oxidative) component of the response. The ratio of the forward and backward peak currents ($I_{p,f}/I_{p,b}$) and the difference between the potentials of the extremes of the components ($E_{p,f} - E_{p,b}$) change from -1.7 and 4 mV (for $\lambda > 20$) to -1.41 and -4 mV (for $\lambda = 1$, see Fig. 2.24a) and -0.92 and -14 mV (for $\lambda < 0.3$, see Fig. 2.24b). These changes are the consequences of the diminished efficacy of SWV stripping technique on thin mercury films [44–46]. Figure 2.26 shows the dependence of the dimensionless net

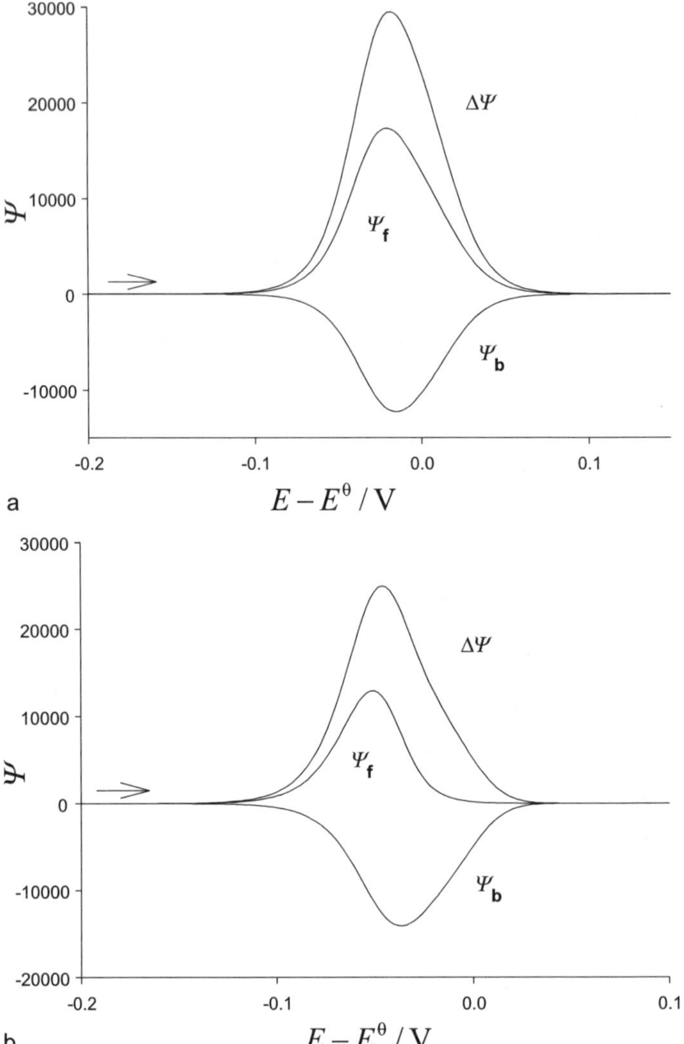

Fig. 2.24a,b Anodic stripping SWV of the reversible reaction (2.25) on the thin mercury film covered rotating disk electrode; $\lambda = 1.118$ (**a**) and 0.2236 (**b**), $n\Delta E = 4$ mV, $nE_{sw} = 50$ mV, $E_{acc} = -0.3$ V, $E_{st} = -0.3$ V, $t_{acc} = 60$ s, $t_{rest} = 10$ s and $\delta = 10^{-3}$ cm

stripping peak currents ($\Delta\Psi_p^* = \Delta I_p/nFAc_{M(Hg)}^* D^{1/2} f^{1/2}$) on the logarithm of the parameter λ. The dependence was calculated by assuming that in (2.25) only metal atoms were initially present in the mercury film, while there were no metal ions in the solution. This means that the preconcentration stage was not considered. The relationship between $\Delta\Psi_p^*$ and $\log\lambda$ is sigmoidal, increasing from $\Delta\Psi_p^* < 0.026$, for $\log\lambda < -1$, to $\Delta\Psi_p^* = 0.7418$, for $\log\lambda > 1.2$. No maximum can be observed. This relationship can be explained by the fact that in thin films the initial amount of metal atoms is much smaller than in thick films. However, more important is that the oxidation of metal atoms in the thin film is much faster than in thick films because in the latter the diffusion is the rate determining step. In very thin films the oxidation may occur in the beginning of the pulse, so that almost no oxidative current can be recorded at the end of the pulse. For this reason the factors of preconcentration, which are very high at thin films, are counterbalanced by the diminished sensitivity of SWV for the accumulated metal atoms in the mercury film electrode. This explains the maximum and the limiting value of $\Delta\Psi_p$ shown in Fig. 2.25a.

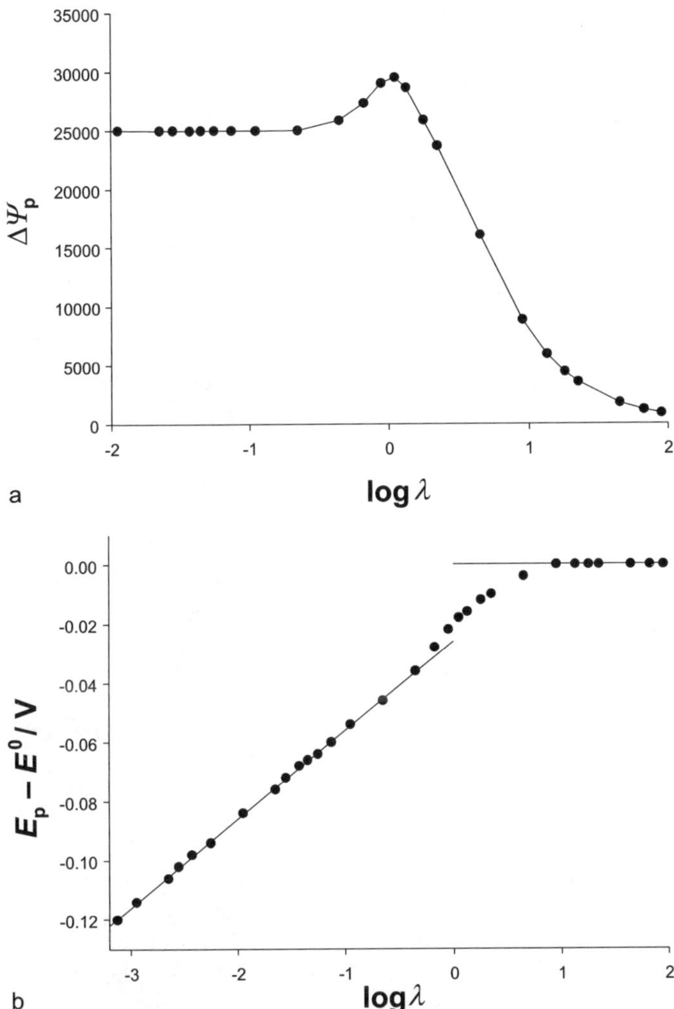

Fig. 2.25a,b Dependence of dimensionless net stripping peak currents (**a**) and the net stripping peak potentials (**b**) on the logarithm of dimensionless film thickness; $n\Delta E = 4\,\mathrm{mV}$, $nE_{sw} = 50\,\mathrm{mV}$, $E_{acc} = -0.3\,\mathrm{V}$, $E_{st} = -0.3\,\mathrm{V}$, $t_{acc} = 60\,\mathrm{s}$, $t_{rest} = 10\,\mathrm{s}$ and $\delta = 10^{-3}\,\mathrm{cm}$

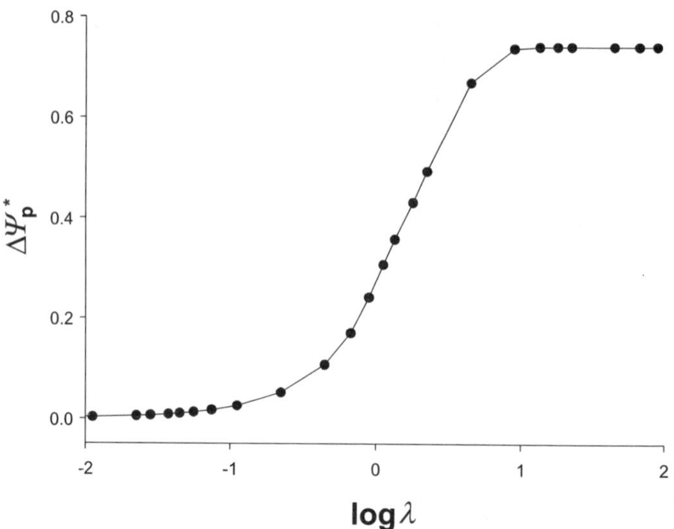

Fig. 2.26 Dependence of dimensionless net stripping peak currents on the logarithm of dimensionless film thickness; $\Delta\Psi_p^* = \Delta I_p/nFAc_{M(Hg)}^* D^{1/2}f^{1/2}$; $n\Delta E = 4\,\mathrm{mV}$, $nE_{sw} = 50\,\mathrm{mV}$, $E_{st} = -0.3\,\mathrm{V}$; no accumulation step was assumed

Relationship between the net stripping peak potential and the logarithm of the parameter λ is shown in Fig. 2.25b. On very thick films ($\lambda > 10$) E_p is independent of frequency, while on very thin films ($\lambda < 0.3$) it is a linear function of the logarithm of frequency, with the slope $\partial E_p/\partial\log f = 2.3RT/2nF$, as in the case of direct measurements (see Fig. 2.23b).

At a very thin film $\Delta\Psi_p^*$ depends linearly on the parameter λ, as can be seen in Fig. 2.27 [45]. The slope of this straight line is: $\Delta\Psi_p^*/\lambda = 0.23$. So, the net stripping peak current is linearly proportional to the frequency. For the parameters in Fig. 2.27, this relationship is:

$$\Delta I_p = 0.23nFAc_{M(Hg)}^* Lf \tag{2.27}$$

Considering (2.26), the net stripping peak current is independent of the film thickness:

$$\Delta I_p = 0.23nFAc^* Dt_{acc}f/\delta \tag{2.28}$$

which is in agreement with the results in Fig. 2.25a, for $\lambda < 0.2$.

The effects of mercury film electrode morphology in the anodic stripping SWV of electrochemically reversible and quasi-reversible processes were investigated experimentally [47–51]. Mercury electroplated onto solid electrodes can take the form of either a uniform thin film or an assembly of microdroplets, which depends on the substrate [51]. At low square-wave frequencies the relationship between the net peak current and the frequency can be described by the theory developed for the thin-film electrode because the diffusion layers at the surface of microdroplets are overlapped and the mass transfer can be approximated by the planar diffusion model [47, 48].

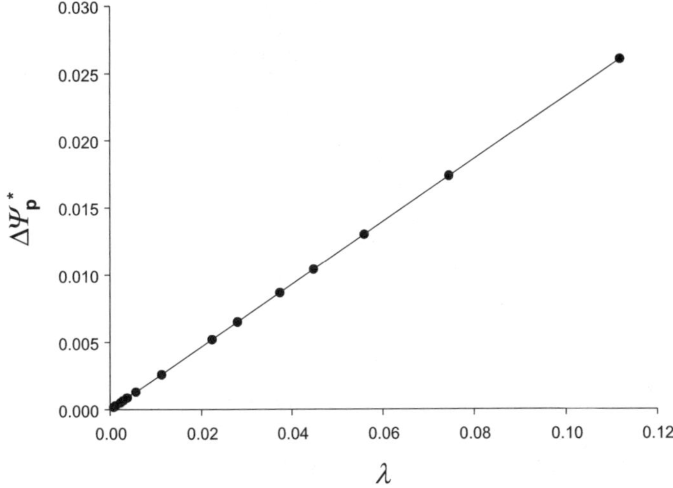

Fig. 2.27 Dependence of dimensionless net stripping peak currents on the dimensionless film thickness; $n\Delta E = 4\,\mathrm{mV}$, $nE_{\mathrm{sw}} = 50\,\mathrm{mV}$, $E_{\mathrm{st}} = -0.3\,\mathrm{V}$; no accumulation step was assumed

However, at the highest frequencies the diffusion layers are much thinner, the micro-droplets tend to behave as independent hemispherical microelectrodes and the response differs significantly from the theory of the thin-film electrode [49].

2.4 Chemical Reactions Coupled to Electrode Reactions

In this chapter electrode mechanisms in which the electrode reaction is coupled with homogeneous chemical reactions are considered. The chemical step, preceeding or following the electrode reaction, is assumed to be of pseudo first order. The theoretical considerations are restricted to the semi-infinite planar diffusion model, which is valid for electrode mechanisms at a macroscopic planar electrode. The numerical solutions are derived by the step function method [52] described in Chap. 1.2. The theoretical background for these electrode mechanisms has been initially provided by the Osteryoungs et al. [15, 53–55]. Latter on, Compton et al. analyzed these types of electrode processes providing a new methodology for numerical simulations based on a backward implicit method for the case of semi-infinite diffusion at a planar [56] and a channel electrode [57]. Rudolph developed a simulation procedure based on the finite difference method adopted for DigSim simulation program, analyzing first and second-order chemical reactions [58]. Molina derived a general analytical solution for catalytic mechanism corresponding to any multi-pulse potential techniques, thus also valid for SWV [59]. Garay and Lovrić analyzed the effect of the kinetics of the electrode reaction in the case of CE and EC reaction schemes for semi-infinite diffusion at a planar electrode [60]. Fatouros et al. also studied theoretically the CE mechanism when the electrode reaction is either reversible or totally irreversible, applying forward and reverse potential scans in the SW voltammetric experiment [24]. The same authors developed a model for a complex triangular catalytic reaction scheme, a sort of combination between a CE and EC' catalytic electrode mechanism [61]. The theory for CE [62], EC [63], and EC' catalytic mechanism [64] at a spherical microelectrode is also available. The experimentally oriented studies in which SWV has been applied to study the coupled chemical reactions are summarized in [65–72].

2.4.1 CE Mechanism

In the CE electrode mechanism the electroactive reactant is produced by means of a preceding homogeneous chemical reaction [15, 55, 60]. Assuming an oxidative mechanism, the simplest form of the CE scheme is as follows:

$$Y \underset{k_b}{\overset{k_f}{\rightleftharpoons}} R \tag{2.29}$$

$$R \rightleftharpoons O + n e^- \tag{2.30}$$

For the sake of simplicity, hereafter the charge of the species is omitted. The preceding chemical reaction is assumed to be a chemically reversible process attributed with first-order forward k_f (s^{-1}) and backward k_b (s^{-1}) rate constants. In the real experimental systems, the forward chemical reaction is most frequently a second-order process:

$$X + Y \underset{k_b}{\overset{k_{f,r}}{\rightleftharpoons}} R \tag{2.31}$$

where $k_{f,r}$ ($mol^{-1} L s^{-1}$) is the real second-order rate constant and X is a certain reactant present in a large excess. The concentration of X is kept constant in the course of the voltammetric experiment. Hence, reaction (2.31) can be treated as being of pseudo first-order characterized by the forward rate constant $k_f = k_{f,r} c_X$. This has an advantage in the experimental analysis, as k_f can be tuned by adjusting the concentration of X.

To take into account the chemical transformation of R, the common diffusion equation is modified as follows:

$$\frac{\partial c_R}{\partial t} = D \frac{\partial^2 c_R}{\partial x^2} + k_f c_Y - k_b c_R \tag{2.32}$$

This equation can be solved in combination with the differential equation describing the mass transport and chemical transformation of Y:

$$\frac{\partial c_Y}{\partial t} = D \frac{\partial^2 c_Y}{\partial x^2} - k_f c_Y + k_b c_R \tag{2.33}$$

For the sake of simplicity, a common diffusion coefficient D is assumed for all species. The mass transport of the O form is described by the common diffusion equation (1.2).

At the beginning of the voltammetric experiment the chemical reaction (2.29) is in equilibrium, characterized by the equilibrium constant K. The latter is the most important thermodynamic parameter of the system, related to the rate constants by $K = \frac{k_f}{k_b}$. Before the voltammetric experiment, the bulk concentrations of Y (c_Y^*) and R (c_R^*) are dictated by the equilibrium constant K and the analytical (total) concentration of the Y (c^*) as follows: $c_Y^* + c_R^* = c^*$ and $K = \frac{c_R^*}{c_Y^*}$. Hence, the experimental conditions prior to the voltammetric experiment are represented by the following initial conditions:

$$t = 0, \ x \geq 0: c_Y = c_Y^* ; \quad c_R = c_R^* ; \quad c_Y^* + c_R^* = c^* ; \quad c_O = 0 ; \quad K = \frac{c_R^*}{c_Y^*} \quad (2.34)$$

In the course of the voltammetric experiment, the conditions in the bulk of the solution remain unaltered. Hence:

$$t > 0, \quad x \to \infty: \quad c_Y \to c_Y^* ; \quad c_R \to c_R^* ; \quad c_Y^* + c_R^* \to c^* ; \quad c_O \to 0 \quad (2.35)$$

At the electrode surface, the fluxes of R and O produce current, whereas the flux of Y at the electrode surface is zero, due to its electrochemical inactivity. Hence, the following boundary conditions hold at the electrode surface:

$$t > 0, \quad x = 0: \quad D\left(\frac{\partial c_R}{\partial x}\right) = -D\left(\frac{\partial c_O}{\partial x}\right) = \frac{I}{nFA} ; \quad \left(\frac{\partial c_Y}{\partial x}\right) = 0 \quad (2.36)$$

The complete mathematical procedure for solving (2.32) is given by Smith [73]. The solution for the surface concentration of R is:

$$(c_R)_{x=0} = c_R^* - \frac{1}{1+K} \int_0^t \frac{I(\tau)}{nFA\sqrt{D}} \frac{e^{-k(t-\tau)}}{\sqrt{\pi(t-\tau)}} \, d\tau - \frac{K}{1+K} \int_0^t \frac{I(\tau)}{nFA\sqrt{D}} \frac{1}{\sqrt{\pi(t-\tau)}} \, d\tau$$

$$(2.37)$$

Here $k = k_f + k_b$ is the kinetic parameter representing the overall kinetics of the chemical step (2.29). The solution for the surface concentration of O is:

$$(c_O)_{x=0} = \int_0^t \frac{I(\tau)}{nFA\sqrt{D}} \frac{1}{\sqrt{\pi(t-\tau)}} \, d\tau \quad (2.38)$$

When the electrode reaction (2.30) is electrochemically reversible, (2.37) and (2.38) are combined with the Nernst equation (1.8) yielding an integral equation that relates the current with time and the electrode potential. The numerical solution derived by the step function method [52] is given by the following recursive formulae:

$$\Psi_m = \frac{\frac{K}{K+1} - \left(\frac{2K}{\sqrt{50\pi}(K+1)} + \frac{2e^{-\varphi m}}{\sqrt{50\pi}}\right) \sum_{j=1}^{m-1} \Psi_j S_{m-j+1} - \frac{1}{\sqrt{\varepsilon}(K+1)} \sum_{j=1}^{m-1} \Psi_j M_{m-j+1}}{\frac{2K}{\sqrt{50\pi}(K+1)} + \frac{M_1}{\sqrt{\varepsilon}(K+1)} + \frac{2e^{-\varphi m}}{\sqrt{50\pi}}}$$

$$(2.39)$$

where the numerical integration parameters are:

$$S_m = \sqrt{m} - \sqrt{m-1} \tag{2.40}$$

and

$$M_m = \mathrm{erf}\left(\sqrt{\frac{\varepsilon\, m}{50}}\right) - \mathrm{erf}\left(\sqrt{\frac{\varepsilon\,(m-1)}{50}}\right), \tag{2.41}$$

where m is the serial number of the time intervals, as explained in Sect. 1.2. It has to be emphasized that the dimensionless current is defined through the total concentration c^*, as $\Psi = \frac{I}{nFAc^*\sqrt{Df}}$. In addition, ε is the critical dimensionless chemical kinetic parameter that relates the kinetic parameter k with the time window of the voltammetric experiment f, defined as $\varepsilon = \frac{k}{f}$. φ is the dimensionless potential defined as in 1.9.

The overall effect of the preceding chemical reaction on the voltammetric response of a reversible electrode reaction is determined by the thermodynamic parameter K and the dimensionless kinetic parameter ε. The equilibrium constant K controls mainly the amount of the electroactive reactant R produced prior to the voltammetric experiment. K also controls the production of R during the experiment when the preceding chemical reaction is sufficiently fast to permit the chemical equilibrium to be achieved on a time scale of the potential pulses. The dimensionless kinetic parameter ε is a measure for the production of R in the course of the voltammetric experiment. The dimensionless chemical kinetic parameter ε can be also understood as a quantitative measure for the rate of reestablishing the chemical equilibrium (2.29) that is misbalanced by proceeding of the electrode reaction. From the definition of ε follows that the kinetic affect of the preceding chemical reaction depends on the rate of the chemical reaction and duration of the potential pulses.

The properties of the response, in particular $\Delta\Psi_\mathrm{p}$ and E_p, vary markedly over different ranges of K and ε. Figure 2.28 shows the variation of $\Delta\Psi_\mathrm{p}$ with ε for different equilibrium constants. By increasing of both K and ε, the preceding chemical reaction produces more electroactive material and the peak current increases. When K and ε are large enough, i.e., for $(\log(K) \geq 2, \log(\varepsilon) \geq 0.2)$, one observes the unperturbed voltammogram for the reversible oxidation of R. When the preceding reaction is slow enough, or the frequency of the potential modulation is high enough $(\log(\varepsilon) < -1)$, the peak current is controlled by the thermodynamic parameter K only. In other words, $\Delta\Psi_\mathrm{p}$ depends on the amount of electroactive material formed by the chemical reaction prior to the electrode process, and the additional production of R during the experiment is insignificant. This situation corresponds to the lower plateau of all curves presented in Fig. 2.28. The rising portion of sigmoid curves represents a typical "kinetic current", reflecting the chemical production of R with the rate dictated by the preceding chemical reaction on the time scale of potential pulses. When the preceding chemical reaction is fast enough, or the frequency of the modulation is low enough, $(\log(\varepsilon) > 3)$, the peak current again depends on K only. As R is consumed during the electrode reaction, the equilibrium of the preceding chemical reaction is shifted toward R, until reaching a new equilibrium, the position of which is dictated by the equilibrium constant K. This situation corresponds to the upper plateau of the sigmoidal curves in Fig. 2.28.

Although it is difficult to generalize the dependence of the peak potential on ε, in general, for an oxidative electrode mechanism, the position of the peak shifts to positive potentials by increasing the rate of the preceding chemical reaction. At the same time, the half-peak width is largely insensitive to the chemical reaction. If $\log(K) \leq -2$, E_p vs. $\log(\varepsilon)$ is a linear function with a slope of about 30 mV.

When the electrode reaction (2.30) is quasireversible, (2.37) and (2.38) are combined with the Butler–Volmer kinetic equation (2.42) [60]:

$$\frac{I}{nFA} = k_s \exp(\alpha_a \varphi)\left[(c_R)_{x=0} - \exp(-\varphi)(c_O)_{x=0}\right] \tag{2.42}$$

where α_a is the anodic electron-transfer coefficient. The numerical solution for quasireversible case reads:

$$\Psi_m = \frac{\kappa e^{\alpha_a \varphi_m}\left[\frac{K}{K+1} - \left(\frac{2K}{\sqrt{50\pi}(K+1)} + \frac{2e^{-\varphi_m}}{\sqrt{50\pi}}\right)\sum_{j=1}^{m-1}\Psi_j S_{m-j+1} - \frac{1}{\sqrt{\varepsilon}(K+1)}\sum_{j=1}^{m-1}\Psi_j M_{m-j+1}\right]}{1 - \kappa e^{\alpha_a \varphi_m}\left[-\frac{2K}{\sqrt{50\pi}(K+1)} - \frac{M_1}{\sqrt{\varepsilon}(K+1)} - \frac{2e^{-\varphi_m}}{\sqrt{50\pi}}\right]} \tag{2.43}$$

In this case, besides the thermodynamic and kinetic parameters of the preceding chemical reaction, the response depends on the kinetics of the electrode reaction represented by the electrode kinetic parameter $\kappa = \frac{k_s}{\sqrt{Df}}$ (see Sect. 2.1.2) [60]. Figure 2.29 shows the variation of $\Delta\Psi_p$ with ε for various κ. It is obvious that there is

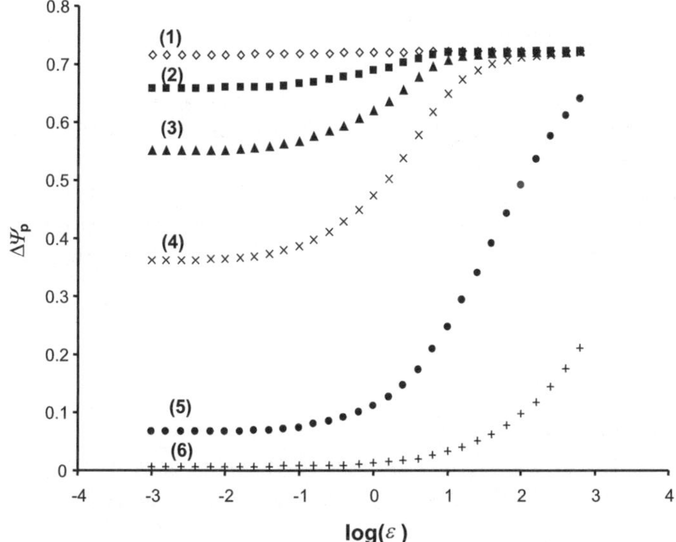

Fig. 2.28 Reversible electrode reaction. $\Delta\Psi_p$ as a function of $\log(\varepsilon)$ for $\log(K) = 2$ (*1*); 1 (*2*); 0.5 (*3*); 0 (*4*); -1 (*5*) and -2 (*6*). The conditions are: $E_{sw} = 50$ mV, $\Delta E = 10$ mV

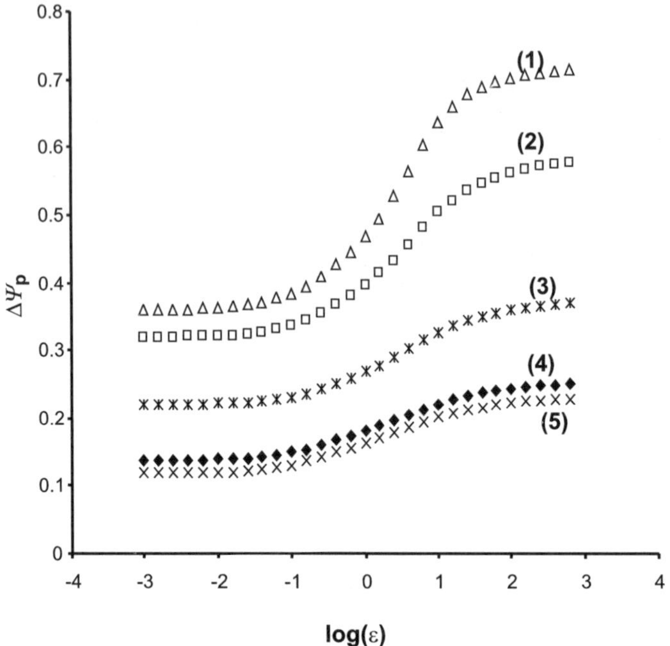

Fig. 2.29 Quasireversible electrode reaction. $\Delta\Psi_p$ as a function of $\log(\varepsilon)$ for $\log(\kappa)=1$ (*1*); 0 (*2*); -0.5 (*3*); -1 (*4*) and -1.5 (*5*). The conditions are: $K=1$, $\alpha_a=0.5$, $nE_{sw}=50\,\text{mV}$, $\Delta E=10\,\text{mV}$

a synergetic influence of the chemical and electrode kinetics. For the quasireversible electrode reaction care must be taken in the analysis of the system in terms of the frequency, as the latter affects simultaneously both the electrode and the chemical kinetic parameters. Garay et al. [60] indicated that if $k>10^7\,\text{s}^{-1}$, the chemical reaction can be considered as totally reversible, and the variation of response by altering the frequency can be attributed to the electrode kinetics only. On the other hand, for $k_s>3\,\text{cm\,s}^{-1}$ the frequency affects only the chemical apparent kinetics.

Experimental studies of CE mechanisms with SWV are scarce. Santos et al. [65] studied two experimental systems, i.e., the reduction of Cd^{2+} ion in the presence of nitrilotriacetic acid (NTA) and aspartic acid (ASP). For the first experimental system, the preceding chemical reaction is described by the scheme:

$$CdNTA^- + H^+ \underset{k_b}{\overset{k_{f,r}}{\rightleftharpoons}} Cd^{2+} + HNTA^{2-} \tag{2.44}$$

The real forward rate constant was estimated to be $k_{f,r}=3.9\times10^5\,\text{mol}^{-1}\text{L\,s}^{-1}$ at $T=25\,°\text{C}$ in $1\,\text{mol/L}$ KNO_3 as a supporting electrolyte. This value is in good agreement with the values measured with cyclic voltammetry and differential pulse polarography [65].

In the system Cd^{2+} and ASP, 1:1 and 1:2 type of complexes are formed. As the dissociation of the 1:2 complex is very fast, the rate determining step is the dissociation of the 1:1 complex:

$$\text{CdASP} \underset{k_b}{\overset{k_{f,r}}{\rightleftarrows}} \text{Cd}^{2+} + \text{ASP}^{2-} \tag{2.45}$$

The forward rate constant of the above reaction was found to be $k_{f,r} = 1.5 \times 10^4\,\text{mol}^{-1}\,\text{L}\,\text{s}^{-1}$ in 0.7 mol/L NaClO$_4$.

Using a glassy carbon electrode modified with a mercury film, Weber et al. [66] measured the association and dissociation rate constants for the complex formed between Pb^{2+} and the 18-crown-6 ether. It was found that Pb^{2+} forms a complex with 18-crown-6 with a stoichiometry of 1:1 in both nitrate and perchlorate media. The formation constant, $k_{f,r}$, for the nitrate and perchlorate systems are $(3.82 \pm 0.89) \times 10^7$ and $(5.92 \pm 1.97) \times 10^6\,\text{mol}^{-1}\,\text{L}\,\text{s}^{-1}$, respectively. The dissociation rate constants, k_b, are $(2.83 \pm 0.66) \times 10^3$ with nitrate and $(2.64 \pm 0.88) \times 10^2\,\text{s}^{-1}$ with perchlorate as counter ion. In addition, the binding of Pb^{2+} with benzo-18-crown-6 embedded in a polymerized crystalline colloidal array hydrogel has been also analyzed [67].

A CE reaction mechanism was also observed for the reduction of iodine at the three-phase electrodes [72]. This electrode system, described in more detail in Chap. 4, consists of a paraffin impregnated graphite electrode at the surface of which a single macroscopic droplet of a nitrobenzene solution containing dissolved iodine is attached. This modified electrode was immersed into an aqueous electrolyte containing Cl$^-$ ions to study the reduction of iodine. It was found that the iodine reduction in the nitrobenzene phase is accompanied by expulsion of chloride ions from nitrobenzene to the aqueous electrolyte. These chloride ions enter the nitrobenzene phase in a preceding reactive partition between aqueous and organic phase, supported by formation of I$_2$Cl$^-$ ions. The overall electrode mechanism follows a CE reaction scheme, where the preceding chemical step is represented by the formation of I$_2$Cl$^-$, whereas the electrode reaction is the reduction of I$_2$ (or I$_2$Cl$^-$) in the organic phase.

2.4.2 EC Mechanism

The simplest case of an EC mechanism is represented by the scheme [15, 55, 60]:

$$\text{R} \rightleftarrows \text{O} + n\text{e}^- \tag{2.46}$$

$$\text{O} \underset{k_b}{\overset{k_f}{\rightleftarrows}} \text{Y} \tag{2.47}$$

The electrode reaction (2.46) is followed by a first-order homogeneous chemical reaction (2.47), in which the product of the electrode reaction O is converted to a final electroinactive product Y. By analogy with the CE mechanism, the chemical step can proceed as:

$$\text{O} + \text{X} \underset{k_R}{\overset{k_{f,r}}{\rightleftarrows}} \text{Y} \tag{2.48}$$

associated with the rate constant $k_f = k_{f,r}c_X$. The meaning of all parameters is equivalent as for CE mechanism. The procedure for modeling of the mass transport of R, as an electroactive reactant is described in Sect. 1.2, whereas the diffusion O form accompanied by chemical transformation is analogous as in the case of CE mechanism and is given by the following model:

$$\frac{\partial c_O}{\partial t} = D\frac{\partial^2 c_O}{\partial x^2} - k_f c_O + k_b c_Y \tag{2.49}$$

$$\frac{\partial c_Y}{\partial t} = D\frac{\partial^2 c_Y}{\partial x^2} + k_f c_O - k_b c_Y \tag{2.50}$$

$$t = 0, \quad x \geq 0: \quad c_R = c_R^*, \quad c_O = c_Y = 0 \tag{2.51}$$

$$t > 0, \quad x \to \infty: \quad c_R \to c_R^*; \quad c_O \to 0; \quad c_Y \to 0 \tag{2.52}$$

$$t > 0, \quad x = 0: \quad D\left(\frac{\partial c_R}{\partial x}\right) = -D\left(\frac{\partial c_O}{\partial x}\right) = \frac{I}{nFA}; \quad D\left(\frac{\partial c_Y}{\partial x}\right) = 0 \tag{2.53}$$

The solutions are:

$$(c_R)_{x=0} = c_R^* - \int_0^t \frac{I(\tau)}{nFA\sqrt{D}} \frac{1}{\sqrt{\pi(t-\tau)}} \, d\tau \tag{2.54}$$

$$(c_O)_{x=0} = \frac{1}{1+K} \int_0^t \frac{I(\tau)}{nFA\sqrt{D}} \frac{1}{\sqrt{\pi(t-\tau)}} \, d\tau + \frac{K}{1+K} \int_0^t \frac{I(\tau)}{nFA\sqrt{D}} \frac{e^{-k(t-\tau)}}{\sqrt{\pi(t-\tau)}} \, d\tau \tag{2.55}$$

The meaning of all symbols is equivalent as for CE mechanism (Sect. 2.4.1). Combining (2.54) and (2.55) with the Nernst equation (1.8) yields an integral equation, as a general solution for a reversible EC mechanism. The numerical solution reads:

$$\Psi_m = \frac{-\frac{2}{\sqrt{50\pi}}\left(\frac{1}{1+K} + e^{\varphi_m}\right)\sum_{j=1}^{m-1} \Psi_j S_{m-j+1} - \frac{K}{\sqrt{\varepsilon(1+K)}}\sum_{j=1}^{m-1} \Psi_j M_{m-j+1} + e^{\varphi_m}}{\frac{2}{\sqrt{50\pi}}\left(\frac{1}{1+K} + e^{\varphi_m}\right) + \frac{K M_1}{\sqrt{\varepsilon(1+K)}}} \tag{2.56}$$

Here the dimensionless current is defined as $\Psi = \dfrac{I}{nFAc_R^*\sqrt{Df}}$.

The voltammetric response depends on the equilibrium constant K and the dimensionless chemical kinetic parameter ε. Figure 2.30 illustrates variation of $\Delta\Psi_p$ with these two parameters. The dependence $\Delta\Psi_p$ vs. $\log(\varepsilon)$, can be divided into three distinct regions. The first one corresponds to the very low observed kinetics of the chemical reaction, i.e., $\log(\varepsilon) < -2$, which is represented by the first plateau of curves in Fig. 2.30. Under such conditions, the voltammetric response is independent of K, since the loss of the electroactive material on the time scale of the experiment is insignificant. The second region, $-2 < \log(\varepsilon) < 4$, is represented by a parabolic dependence characterized by a pronounced minimum. The descending part of the parabola arises from the conversion of the electroactive material to the final inactive product, which is predominantly controlled by the rate of the forward chemical reaction. However, after reaching a minimum value, the peak current starts to increase by an increase of ε. In the ascending part of the parabola, the effect of the reverse chemical reaction on the time scale of the reverse (reduction) potential pulses becomes predominant. This means that the backward chemical reaction is sufficiently fast to re-supply the electroactive form O, which is consumed by the electrode reaction during the reductive backward pulses. It is important to note that the position of the minimum in Fig. 2.30 depends on K, being shifted toward larger values of ε by increasing K. This is a consequence of the fact that the enhancing of the thermodynamic parameter K favors the production of the electroinactive product. Consequently, higher rates of the chemical reaction are required to re-supply the electroactive material on the time scale of the backward (reduction) pulses.

Similar to the CE mechanism, the half-peak width of the net response for the EC mechanism is largely insensitive to the chemical reaction. The variation of the peak potential with K and ε cannot be generalized over a wide range of values. For an oxidative electrode mechanism, an increase of ε causes generally a shift of the peak potential toward more negative values. However, for a given K, the peak potential reaches a certain limiting value for sufficiently large ε.

If the electrode reaction is quasireversible, (2.54) and (2.55) are combined with the kinetic equation (2.42). The numerical solution reads:

$$\Psi_m = \frac{\kappa\, e^{\alpha_a \varphi_m}\left[1 - \frac{2}{\sqrt{50\pi}}\left(1 + \frac{e^{-\varphi_m}}{1+K}\right)\sum_{j=1}^{m-1}\Psi_j S_{m-j+1} - \frac{K e^{-\varphi_m}}{(1+K)\sqrt{\varepsilon}}\sum_{j=1}^{m-1}\Psi_j M_{m-j+1}\right]}{1 - \kappa\, e^{\alpha_a \varphi_m}\left[-\frac{2}{\sqrt{50\pi}}\left(1 + \frac{e^{-\varphi_m}}{1+K}\right) - \frac{e^{-\varphi_m} K M_1}{(1+K)\sqrt{\varepsilon}}\right]}$$

$$(2.57)$$

where all symbols have the same meaning as for the CE mechanism.

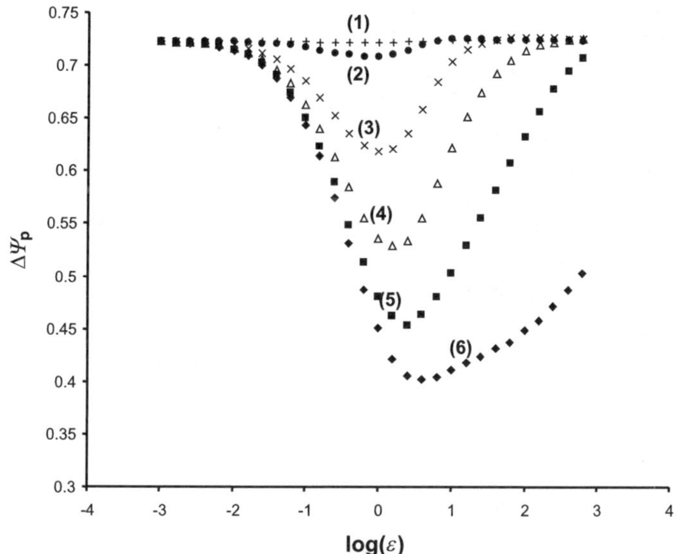

Fig. 2.30 Reversible electrode reaction. $\Delta\Psi_p$ as a function of $\log(\varepsilon)$ for $\log(K) = -2$ (*1*); -1 (*2*); 0 (*3*); 0.5 (*4*); 1 (*5*) and 2 (*6*). The conditions are: $nE_{sw} = 50\,\mathrm{mV}$, $\Delta E = 10\,\mathrm{mV}$

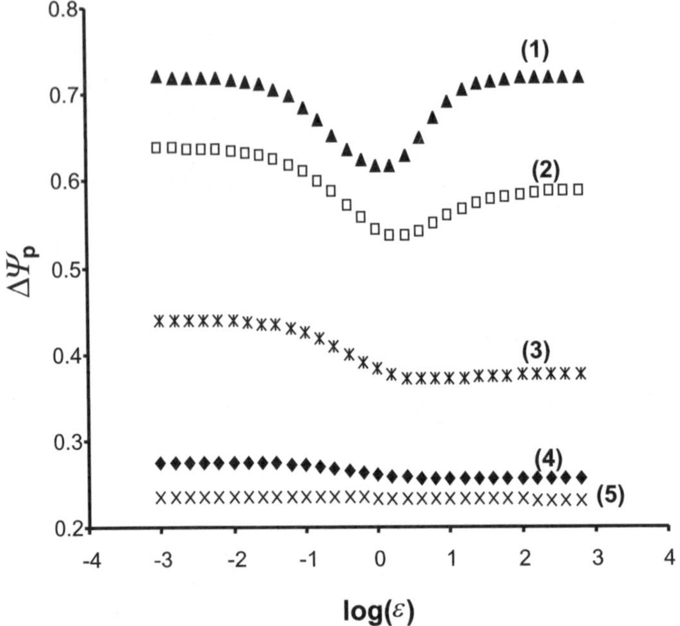

Fig. 2.31 Quasireversible electrode reaction. $\Delta\Psi_p$ as a function of $\log(\varepsilon)$ for $\log(\kappa) = 1$ (*1*); 0 (*2*); -0.5 (*3*); -1 (*4*) and -1.5 (*5*). The conditions are: $K = 1$, $\alpha_a = 0.5$, $nE_{sw} = 50\,\mathrm{mV}$, $\Delta E = 10\,\mathrm{mV}$

The variation of the peak current with the electrode kinetic parameter κ and chemical kinetic parameter ε is shown in Fig. 2.31. When the quasireversible electrode reaction is fast (curves 1 and 2 in Fig. 2.31) the dependence is similar as for the reversible case and characterized by a pronounced minimum. If the electrode reaction is rather slow (curves 3–5), the dependence $\Delta\Psi_p$ vs. $\log(\varepsilon)$ transforms into a sigmoidal curve. Although the backward chemical reaction is sufficiently fast to re-supply the electroactive material on the time scale of the reverse (reduction) potential pulses, the reuse of the electroactive form is prevented due to the very low kinetics of the electrode reaction. This situation corresponds to the lower plateau of curves 3–5 in Fig. 2.31.

The validity of the theoretical predictions is yet not experimentally rigorously confirmed by a model experimental system, although the theory has a safe background in the theory and experiments of similar potential pulse techniques as well as cyclic staircase voltammetry.

2.4.3 ECE Mechanism

The ECE mechanism [54] unifies the previously elaborated EC and CE mechanisms. It is represented by the following scheme:

$$R_1 \rightleftarrows O_1 + n_1 e^- \tag{2.58}$$

$$O_1 \xrightarrow{k_f} R_2 \tag{2.59}$$

$$R_2 \rightleftarrows O_2 + n_2 e^- \tag{2.60}$$

In the ECE mechanism, the electrode reaction (2.58) is followed by a homogeneous irreversible chemical reaction (2.59) that generates a new electroactive reactant undergoing further electrochemical transformation through the second electrode reaction (2.60). Although this reaction scheme is rather complex, it is a common reaction pathway in organic electrochemistry. For the sake of simplicity, it is assumed that both electrode reactions are reversible, characterized with standard potentials E_1^θ and E_2^θ, whereas the chemical step is totally irreversible attributed with a first-order rate constant k_f (s^{-1}). Mathematical modeling of the diffusion mass transport of the initial electroactive reactant R_1, as well as the final electroactive product O_2 follows the common procedure described in Sect. 1.2. Diffusion of O_1 and R_2 is described by the following differential equations:

$$\frac{\partial c_{O_1}}{\partial t} = D \frac{\partial^2 c_{O_1}}{\partial x^2} - k_f c_{O_1} \tag{2.61}$$

$$\frac{\partial c_{R_2}}{\partial t} = D \frac{\partial^2 c_{R_2}}{\partial x^2} + k_f c_{O_1} \tag{2.62}$$

The initial and boundary conditions are:

$$t = 0, \quad x \geq 0: \qquad c_{R_1} = c_{R_1}^*, \quad c_{O_1} = c_{R_2} = c_{O_2} = 0 \tag{2.63}$$

$$t > 0, \quad x = 0: \qquad D\left(\frac{\partial c_{R_1}}{\partial x}\right) = -D\left(\frac{\partial c_{O_1}}{\partial x}\right) = \frac{I_1}{n_1 FA} \tag{2.64}$$

$$D\left(\frac{\partial c_{R_2}}{\partial x}\right) = -D\left(\frac{\partial c_{O_2}}{\partial x}\right) = \frac{I_2}{n_2 FA} \tag{2.65}$$

$$t > 0, \quad x = 0: \qquad (c_{O_1})_{x=0} = \exp(\varphi_1)(c_{R_1})_{x=0} \tag{2.66}$$

$$t > 0, \quad x = 0: \qquad (c_{O_2})_{x=0} = \exp(\varphi_2)(c_{R_2})_{x=0} \tag{2.67}$$

Here, I_1 and I_2 are current contributions arising from electrode reactions (2.58) and (2.60), respectively. Of course, in the experiment, only the total current $I = I_1 + I_2$ is observable. $\varphi_1 = \frac{n_1 F}{RT}(E - E_1^\theta)$ and $\varphi_2 = \frac{n_2 F}{RT}(E - E_2^\theta)$ are relative dimensionless potentials for reactions (2.58) and (2.60), respectively. The solutions for surface concentrations of all electroactive species are as follows:

$$(c_{R_1})_{x=0} = c_{R_1}^* - \int_0^t \frac{I_1(\tau)}{n_1 FA\sqrt{D}} \frac{1}{\sqrt{\pi(t - \tau)}} \, d\tau \tag{2.68}$$

$$(c_{O_1})_{x=0} = \int_0^t \frac{I_1(\tau)}{n_1 FA\sqrt{D}} \frac{e^{-k_f(t-\tau)}}{\sqrt{\pi(t-\tau)}} \, d\tau \tag{2.69}$$

$$(c_{R_2})_{x=0} = \int_0^t \frac{I_1(\tau)}{n_1 FA\sqrt{D}} \frac{d\tau}{\sqrt{\pi(t-\tau)}} - \int_0^t \frac{I_1(\tau)}{n_1 FA\sqrt{D}} \frac{e^{-k_f(t-\tau)}}{\sqrt{\pi(t-\tau)}} \, d\tau$$

$$- \int_0^t \frac{I_2(\tau)}{n_2 FA\sqrt{D}} \frac{d\tau}{\sqrt{\pi(t-\tau)}} \tag{2.70}$$

$$(c_{O_2})_{x=0} = \int_0^t \frac{I_2(\tau)}{n_2 FA\sqrt{D}} \frac{d\tau}{\sqrt{\pi(t-\tau)}} \tag{2.71}$$

An integral equation describing the contribution of the first electrode reaction can be easily derived by substituting (2.68) and (2.69) into (2.66). Accordingly, the integral equation corresponding to the second electrode reaction (2.60) can be readily obtained by substituting (2.70) and (2.71) into (2.67). The numerical solutions for the dimensionless current contributions of the two electrode reactions are given as follows:

$$(\Psi_1)_m = \frac{1 - \frac{e^{-(\varphi_1)m}}{\sqrt{\varepsilon}} \sum_{j=1}^{m-1} (\Psi_1)_j M_{m-j+1} - \frac{2}{\sqrt{50\pi}} \sum_{j=1}^{m-1} (\Psi_1)_j S_{m-j+1}}{\frac{2}{\sqrt{50\pi}} + \frac{e^{-(\varphi_1)m} M_1}{\sqrt{\kappa}}} \tag{2.72}$$

$$(\Psi_2)_m = \frac{\frac{1}{\sqrt{\varepsilon}} \sum_{j=1}^{m} (\Psi_1)_j M_{m-j+1} - \frac{2}{\sqrt{50\pi}} \sum_{j=1}^{m} (\Psi_1)_j S_{m-j+1} + \frac{2(1+e^{-(\varphi_2)m})}{\sqrt{50\pi}} \sum_{j=1}^{m-1} (\Psi_2)_j S_{m-j+1}}{-\frac{2}{\sqrt{50\pi}} (1 + e^{-(\varphi_2)m})} \tag{2.73}$$

where $\Psi_1 = \frac{I_1}{n_1 FAc_{R_1}^* \sqrt{Df}}$ and $\Psi_2 = \frac{I_2}{n_2 FAc_{R_1}^* \sqrt{Df}}$. The total dimensionless current is $\Psi = \Psi_1 + \Psi_2$. The integration factors S_m and M_m are defined by (2.40) and (2.41), respectively. Note, that in the present case, the dimensionless chemical kinetic parameter ε is defined as $\varepsilon = \frac{k_f}{f}$.

Depending on the difference between the standard potentials, $\Delta E^\theta = E_2^\theta - E_1^\theta$ and the dimensionless kinetic parameter ε, a large variety of voltammetric profiles can be obtained. A few examples are shown in Figs. 2.32 to 2.34. The panel a of each figure shows separately the contributions of each electrode reaction, whereas the panel b and c depict components of the total response. Figure 2.32 corresponds to the case when $\Delta E^\theta = 0.3$ V and low apparent kinetics of the chemical reaction, i.e., $\log(\varepsilon) = -2$. In this case the second electrochemical step requires higher energy than the first one. For an oxidative mechanism, it means that the second oxidation step undergoes at more positive potentials. The composite net response consists of two well-separated peaks. The second peak arises from the reversible oxidation of R_2, which is generated in situ during the potential scan.

Figure 2.33 shows an intermediate case when the formal potentials of the two redox couples are the same, and the chemical reaction occurs with a moderate rate,

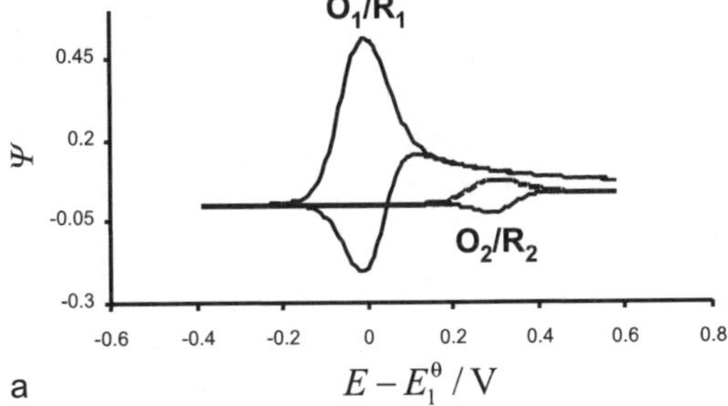

a

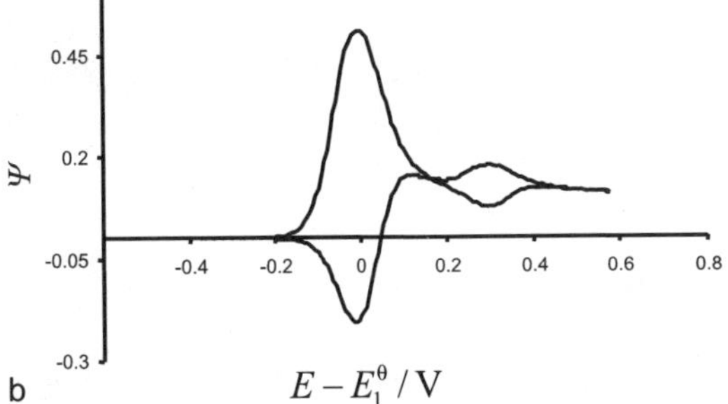

b

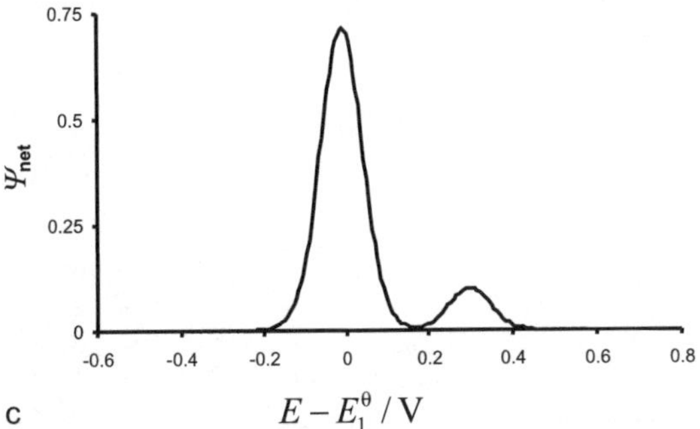

c

Fig. 2.32a–c Theoretical voltammograms simulated for $\log(\varepsilon) = -2$ and $E_2^\theta - E_1^\theta = 0.3$ V. The conditions are: $n_1 = n_2 = 1$, $E_{sw} = 50$ mV, $\Delta E = 10$ mV. (**a**) Contribution from O_1/R_1 and O_2/R_2 couples; (**b**) combined forward and reverse currents; (**c**) net current

$\log(\varepsilon) = 0$. From the panel a, it can be seen that part of the redox form O_2 is re-reduced back to R_2, which causes enhancement of the total net current.

Particularly interesting is the case when the second electrode reaction requires lower energy than the first one, i.e., R_2 is more easily oxidized than R_1. In this case the total response consists of a single peak. The exact shape and position of this peak and its forward and reverse components reflect the relative contributions of the redox couples R_1/O_1 and R_2/O_2 over a narrow range of potentials dictated by the oxidation of R_1. As a consequence, the response due to R_1/O_1 masks the response

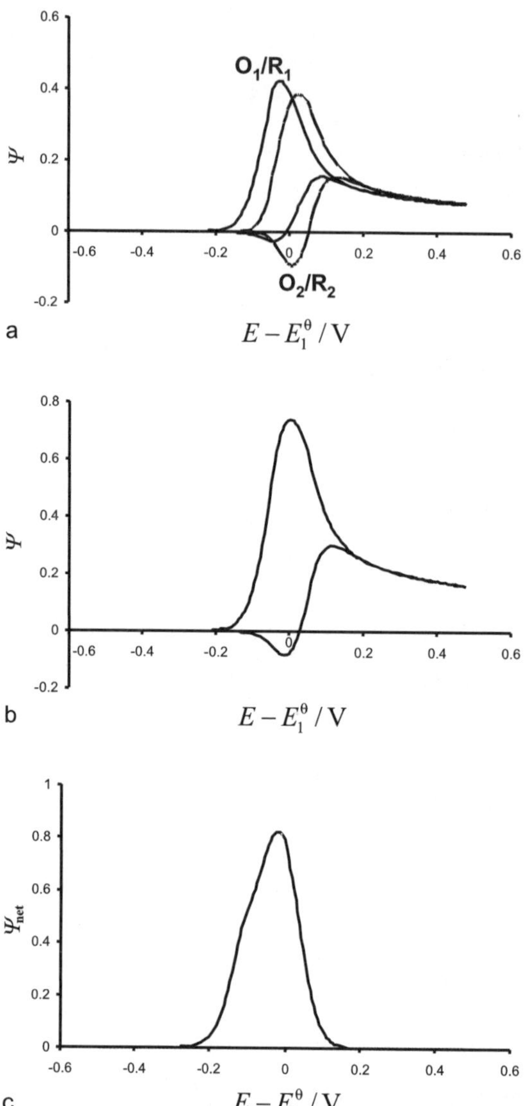

Fig. 2.33a–c Theoretical voltammograms simulated for $\log(\varepsilon) = 0$ and $E_2^\theta = E_1^\theta$. All other notations and conditions are the same as for Fig. 2.32

originating from R$_2$/O$_2$. Figure 2.34 shows voltammetric profiles corresponding to such case ($\Delta E^\theta = -0.3$ V), when apparent kinetics of the chemical reaction is high ($\log(\varepsilon) = 2$). Although both electrode reactions are reversible, no reverse response is seen. As soon as O$_1$ is formed close to the standard potential E_1^θ, it transforms to R$_2$ due to the fast chemical reaction. Thus, it cannot be re-reduced back to R$_1$ and the reverse component of the first electrode reaction vanishes. As soon as R$_2$ is

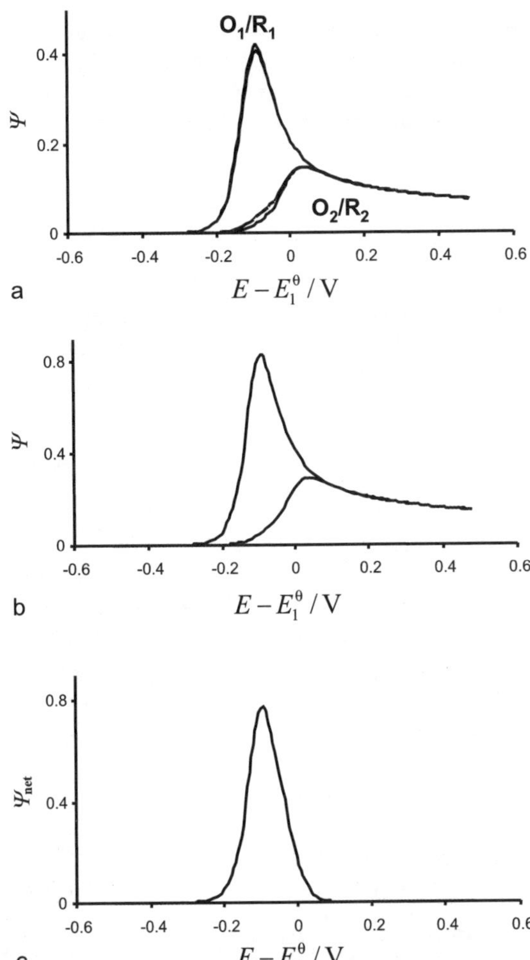

Fig. 2.34a–c Theoretical voltammograms simulated for $\log(\varepsilon) = 2$ and $E_2^\theta - E_1^\theta = -0.3$ V. All other notations and conditions are the same as for Fig. 2.32

formed in situ close to the formal potential of the first redox couple, it is oxidized immediately to O_2 since the potential is far more positive than E_2^θ. However, the reduction of O_2 to R_2 cannot proceed since the potential is not sufficiently negative for this process. As a consequence, the reverse component of the couple R_2/O_2 does not emerge. A useful strategy to detect the reduction of O_2 is to scan the square-wave waveform in negative direction. Starting from potentials more positive than E_1^θ, the O_2 can be formed in situ, which then can be detected by negative ongoing potential scan at potentials close to E_2^θ.

The experimental model used to illustrate the ECE mechanism was the reduction of p-nitrosophenol at a mercury electrode, in which the chemical step is dehydration [54]. The experimental data have been analyzed by best-fitting curve procedure using COOL algorithm. At 25 °C the chemical rate constant measured at $f = 10\,\mathrm{Hz}$ was $0.46\,\mathrm{s}^{-1}$ being in a good agreement with the value of $0.6\,\mathrm{s}^{-1}$ reported by Nicholson and Shain [74]. By measuring the dependence of the chemical rate constant on temperature, the activation energy of $57\,\mathrm{kJ/mol}$ was estimated.

2.4.4 EC' Catalytic Mechanism

We finally consider the EC' catalytic mechanism in which the product of the electrode reaction transforms back to the initial electroactive reactant by means of a pseudo first-order chemical reaction [15, 53, 55]:

$$R \rightleftharpoons O + ne^- \tag{2.74}$$

$$O_1 \xrightarrow{k_c} R_1 \tag{2.75}$$

Note that the chemical step (2.75) is totally irreversible, attributed with a pseudo first-order rate constant k_c (s^{-1}) defined as $k_c = k_{c,r}c_X$, where c_X has the same meaning as for the CE and EC mechanisms (Sect. 2.4.1). Although this is the simplest case of an electrode mechanism involving chemical reaction, it has particular analytical utility [53]. The mass transport of the redox species is described by the following model:

$$\frac{\partial c_R}{\partial t} = D\frac{\partial^2 c_R}{\partial x^2} + k_c c_O \tag{2.76}$$

$$\frac{\partial c_O}{\partial t} = D\frac{\partial^2 c_O}{\partial x^2} - k_c c_O \tag{2.77}$$

$$t = 0, \quad x \geq 0: \quad c_R = c_R^*, \quad c_O = 0 \tag{2.78}$$

$$t > 0, \quad x \to \infty: \quad c_R \to c_R^*; \quad c_O \to 0 \tag{2.79}$$

$$t > 0, \quad x = 0: \quad D\left(\frac{\partial c_R}{\partial x}\right) = -D\left(\frac{\partial c_O}{\partial x}\right) = \frac{I}{nFA} \tag{2.80}$$

The solutions for the surface concentrations of redox species are:

$$(c_R)_{x=0} = c_R^* - \int_0^t \frac{I(\tau)}{nFA\sqrt{D}} \frac{e^{-k_c(t-\tau)}}{\sqrt{\pi(t-\tau)}}\,d\tau \tag{2.81}$$

$$(c_O)_{x=0} = \int_0^t \frac{I(\tau)}{nFA\sqrt{D}} \frac{e^{-k_c(t-\tau)}}{\sqrt{\pi(t-\tau)}}\,d\tau \tag{2.82}$$

By substituting (2.81) and (2.82) into the Nernst equation (1.8), one obtains an integral equation, as a solution for a reversible catalytic mechanism. The numerical solution for the reversible case reads:

$$\Psi_m = \frac{\frac{e^{\varphi_m}\sqrt{\varepsilon}}{1+e^{\varphi_m}} - \sum_{j=1}^{m-1} \Psi_j M_{m-j+1}}{M_1} \tag{2.83}$$

where $\varepsilon = \frac{k_c}{f}$ is dimensionless chemical kinetic parameter and all other symbols have the same meaning as for EC mechanism (Sect. 2.4.2).

For the catalytic electrode mechanism, the total surface concentration of R plus O is conserved throughout the voltammetric experiment. As a consequence, the position and width of the net response are constant over entire range of values of the parameter ε. Figure 2.35 shows that the net peak current increases without limit with ε. This means that the maximal catalytic effect in particular experiment is obtained at lowest frequencies. Figure 2.36 illustrates the effect of the chemical reaction on the shape of the response. For $\log(\varepsilon) < -3$, the response is identical as for the simple reversible reaction (curves 1 in Fig. 2.36). Due to the effect of the chemical reaction which consumes the O species and produces the R form, the reverse component decreases and the forward component enhances correspondingly (curves 2 in Fig. 2.36). When the response is controlled exclusively by the rate of the chemical reaction, both components of the response are sigmoidal curves separated by $2E_{sw}$ on the potential axes. As shown by the inset of Fig. 2.36, it is important to note that the net currents are bell-shaped curves for any observed kinetics of the chemical reaction, with readily measurable peak current and potentials, which is of practical importance in electroanalytical methods based on this electrode mechanism.

If the electrode reaction is quasireversible, (2.81) and (2.82) are combined with the kinetic equation (2.42). The numerical solution for the quasireversible case is:

$$\Psi_m = \frac{\kappa\, e^{\alpha_a \varphi_m}\left[1 - \frac{1+e^{-\varphi_m}}{\sqrt{\varepsilon}}\sum_{j=1}^{m-1}\Psi_j M_{m-j+1}\right]}{1 + \frac{\kappa\, e^{\alpha_a \varphi_m} M_1}{\sqrt{\varepsilon}}\left(1 + e^{-\varphi_m}\right)} \tag{2.84}$$

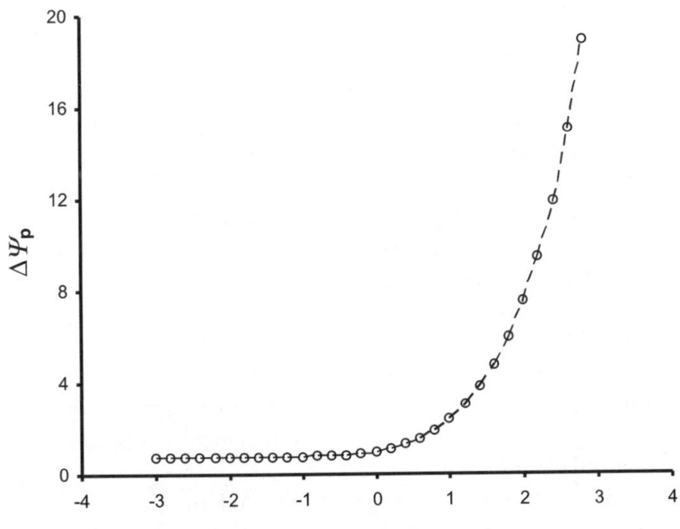

Fig. 2.35 Reversible electrode reaction. $\Delta\Psi_p$ as a function of $\log(\varepsilon)$. The conditions are: $nE_{sw} = 50\,\text{mV}$, $\Delta E = 10\,\text{mV}$

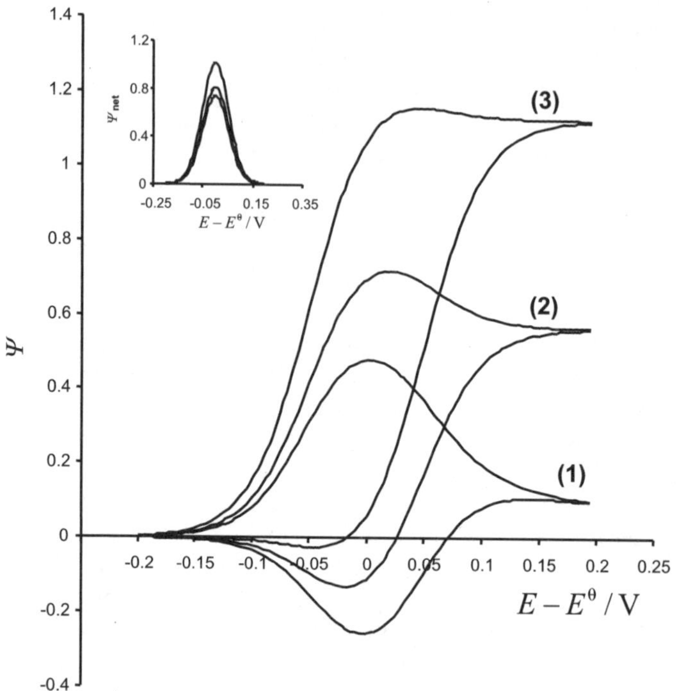

Fig. 2.36 Reversible electrode reaction. The forward and reverse components of the theoretical voltammograms simulated for $\log(\varepsilon) = -3$ (*1*); -0.5 (*2*) and 0.1 (*3*). The *inset* shows the corresponding net voltammograms. The other conditions are the same as for Fig. 2.35

Here, κ has the same meaning as for previous mechanisms. The effects of the dimensionless chemical kinetic parameter ε and the electrode kinetic parameter κ on the net peak currents are depicted in Fig. 2.37. The dependency is identical as for reversible reaction with a difference that the sensitivity of the response to the chemical reaction increases by increasing the apparent reversibility of the electrode reaction. Contrary to the reversible case, the position and width of the net peak are sensitive to the chemical reaction. This effect is more pronounced if the apparent kinetics of the electrode reaction is lower. For an oxidative electrode mechanism, the peak shifts toward more positive potentials by increasing the kinetics of the chemical reaction. At the same time the net peak increases in its width. Generally speaking, the effect observed by increasing the rate of the chemical reaction resembles the effect of decreasing the electrode kinetics. As for previous cases, for quasireversible catalytic mechanism, care must be taken in analysis of the system by varying the frequency of the potential modulation since this parameter affects simultaneously the apparent kinetics of both electrode and chemical reactions.

The theory for catalytic reaction has been verified by studying the reductions of Ti^{4+} in presence of NH_2OH and ClO_3^- and the reduction of Fe^{3+} in presence of NH_2OH. In these studies the mercury electrode has been applied [53]. The properties of the experimental voltammograms confirm the theoretical predic-

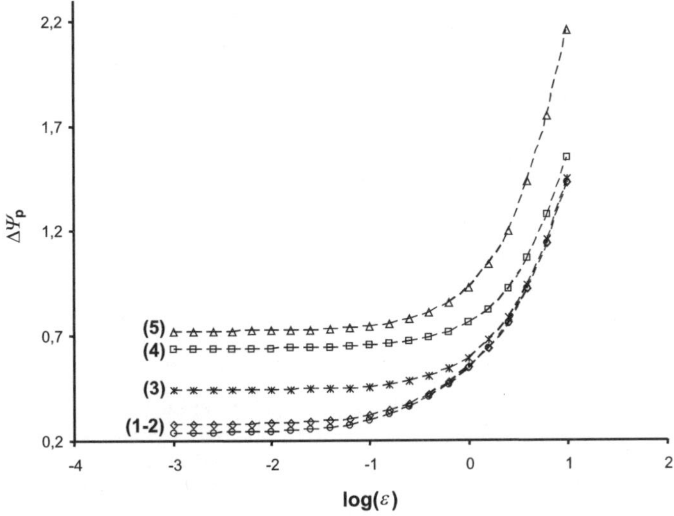

Fig. 2.37 Quasireversible electrode reaction. $\Delta\Psi_p$ as a function of $\log(\varepsilon)$ for $\log(\kappa) = -1.5$ (1); -1 (2); -0.5 (3); 0 (4), and 1 (5). The conditions are: $\alpha_a = 0.5$, $nE_{sw} = 50$ mV, $\Delta E = 10$ mV

tions. Representative voltammograms of the reduction of Ti^{4+} in the presence of NH_2OH and $H_2C_2O_4$ at different frequencies are shown in Fig. 2.38. The symbols refer to the experimental data and the lines represent the theoretical fit. These results illustrate how the chemical parameter ε of the chemical catalytic reaction decreases by increasing the frequency. The rate constants found are: 41 ± 1.02 L mol^{-1} s^{-1} for Ti^{3+}/NH_2OH, $(51.16 \pm 0.33) \times 10^4$ L mol^{-1} s^{-1} for Ti^{3+}/ClO_3^- and 150.8 ± 6.3 L mol^{-1} s^{-1} Fe^{2+}/NH_2OH. These values are in good agreement with literature data obtained by differential pulse polarography [75]. Besides, the authors discussed the advantages of using SWV over other voltammetric method in reference to the analytical utility of the catalytic electrode mechanism.

However, the Ti^{4+}-oxalate system is more complex than considered in the previous study [53]. In order to explain the catalytic electrode mechanism in the presence of chlorate ions, Krulic et al. [61] studied a complex catalytic triangular reaction scheme, a sort of combination of a CE and EC' reaction mechanisms. In an acidic solution containing oxalate ions, Ti^{4+} forms 1:1 and 1:2 oxalate complexes, designated as S_1 and S_2 respectively. The chemical conversion of S_1 to S_2 is a slow chemical process. The equilibrium reaction between the two complexes is primarily determined by pH of the medium and concentration of oxalate ions. By increasing the concentration of oxalate ions, the equilibrium is shifted in favor of the complex S_2. The complexes S_1 and S_2 can be reduced at a mercury electrode to form 1:2 Ti^{3+}-oxalate complex. The electrode reduction of S_2 is a fast and reversible process occurring at more positive potentials than S_1, the reduction of which is a slow, electrochemically irreversible process. In the presence of chlorate ions, the Ti^{3+}-complex is irreversibly oxidized to Ti^{4+}-complexes, thus forming a catalytic reaction scheme. Figure 2.39 shows voltammograms for the reduction of Ti^{4+} in oxalate

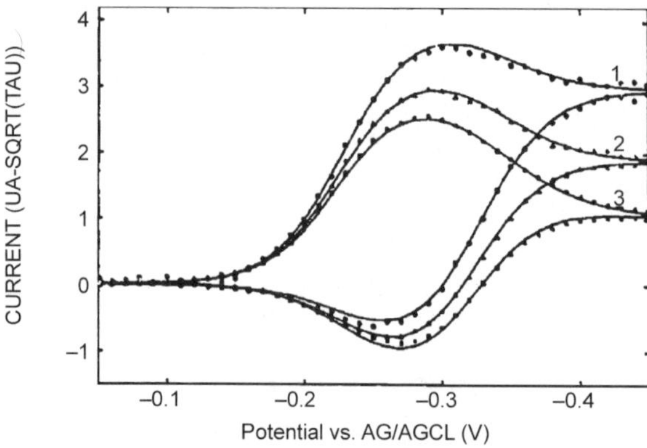

Fig. 2.38 Forward and backward components of the SW response recorded by reduction of 1 mmol/L Ti^{4+} at a mercury electrode in the presence of 0.1 mol/L NH_2OH and 0.2 mol/L $H_2C_2O_4$ for $f = 10$ (*1*); 25 (*2*) and 100 Hz (*3*). *Symbols* are experimental data and *lines* are theoretical fit. The other conditions are: $E_{sw} = 50$ mV, $\Delta E = 10$ mV (reprinted from [53] with permission)

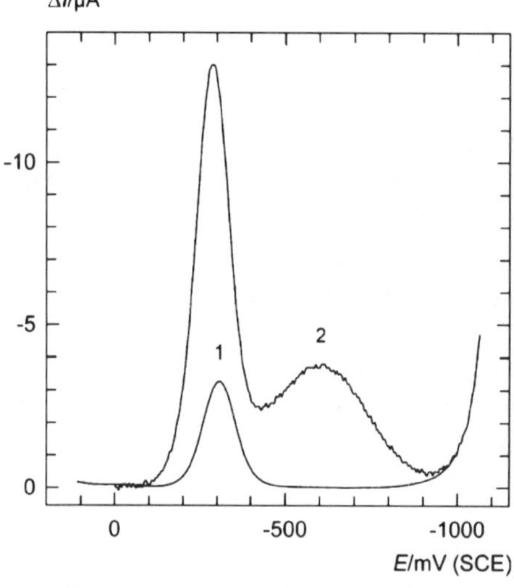

Fig. 2.39 Net SW voltammograms of Ti^{4+} in the absence (*curve 1*) and presence (*curve 2*) of $NaClO_3$. Experimental conditions: $c(Ti^{4+}) = 2 \times 10^{-4}$ mol/L, $c(oxalate) = 0.16$ mol/L, $c(H^+) = 0.027$ mol/L (pH ≈ 1.6), ionic strength 1 mol/L. For *curve 2* $c(NaClO_3) = 0.1$ mol/L, temperature 25 °C, radius of the mercury drop 0.04 cm, $E_{sw} = 80$ mV, $\Delta E = -4$ mV, $f = 50$ Hz (reprinted from [61] with permission)

medium in the absence (curve 1) and presence (curve 2) of chlorate ions. The single net SW peak found in the absence of chlorate ions corresponds to the reduction of S_2 complex, following a CE reaction scheme. The peak due to reduction of S_1 complex is absent due to the negligible low concentration of this complex under the experimental conditions of Fig. 2.39. As inferred from curve 2, chlorate ions caused a significant catalytic effect; moreover, besides the response of the complex S_2, the reduction peak due to the reduction of the complex S_1 emerged at more negative potentials. Obviously, the homogenous oxidation of the complex S_3 by chlorate ions causes an apparent deceleration of the chemical reaction between S_1 and S_2, which leads to the reappearance of the electrochemical response of S_1. Figure 2.40 shows all components of the SW response, revealing a strong catalytic effect that causes steady-state shaped forward and backward components of the SW response, as predicted by the theory (see curve 3 in Fig. 2.36). The catalytic mechanism of Ti^{4+}-oxalate complexes in the presence of oxalate ions can be represented by one of the following schemes:

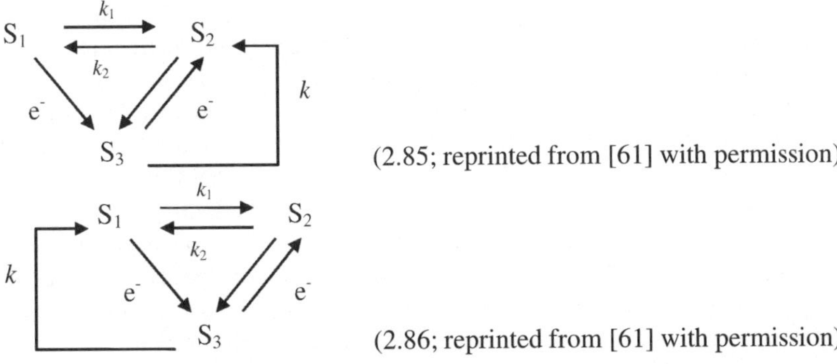

(2.85; reprinted from [61] with permission)

(2.86; reprinted from [61] with permission)

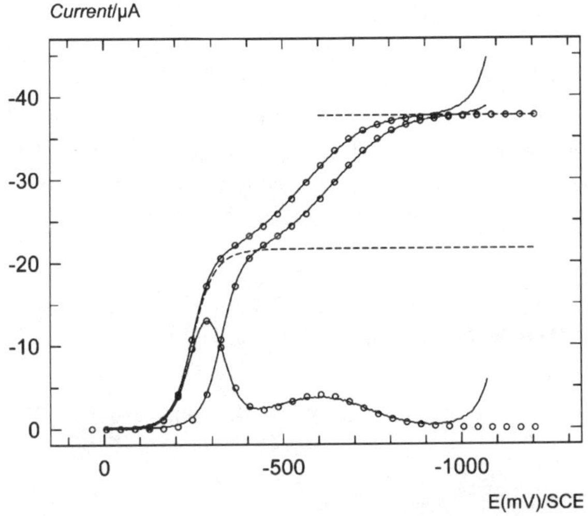

Fig. 2.40 Experimental (*solid lines*) and calculated (*circles*) components of the SW response for the reduction of Ti^{4+} in the presence of $NaClO_4$. All experimental conditions are the same as for Fig. 2.39. The calculations correspond to the mechanism (2.85). The conditions of the calculations are: $D = 3.8 \times 10^{-6}$ cm^2 s^{-1}, $K = 1.4$ ($K = [S_1]/[S_2]$), $k_2 = 1200$ s^{-1}, $k = 2450$ s^{-1}, $E_2^\theta = -305$ mV vs. SCE, $k_{s,2} = 0.37$ cm s^{-1}, $\alpha_{c,2} = 0.5$, $E_1^\theta = -82$ mV vs. SCE, $k_{s,2} = 5 \times 10^{-5}$ cm s^{-1}, $\alpha_{c,1} = 0.33$. All *symbols* as in scheme (2.85) (reprinted from [61] with permission)

The authors have developed theoretical models for both reaction schemes and compared them with the experimental results, concluding that the experimental system follows the reaction scheme (2.85). This was confirmed by the excellent fit between the experimental and calculated data presented in Fig. 2.40.

2.5 Surface Electrode Reactions

Surface electrode reactions are an important class of electrode processes that have been a subject of a long-standing scientific interest in electrochemistry due to their relevance for surface science, electrochemical sensors, characterization of surface-modified electrodes, redox active drugs, proteins, enzymes etc. In surface electrode reactions, the electroactive form is present as a monolayer or sub-monolayer confined to the electrode surface. The immobilization of the electroactive component can be achieved by various means such as adsorption, covalent bonding, self-assembly, Langmuir-Blodgett method, etc. The redox transformation proceeds only within the electroactive film confined to the electrode surface, without significant contribution of the electrode reaction of dissolved species. Hence, in the modeling of surface processes, the diffusion mass transport of electroactive species is not considered, taking into account only the variation of the surface concentrations of electroactive species with time. This simplifies considerably the mathematical procedure in the modeling of the voltammetric experiment.

In the last two decades, significant attention has been paid to the study of surface electrode reactions with SWV and various methodologies have been developed for thermodynamic and kinetic characterization of these reactions. In the following chapter, several types of surface electrode processes are addressed, including simple quasireversible surface electrode reaction [76–84], surface reactions involving lateral interactions between immobilized species [85], surface reactions coupled with chemical reactions [86–89], as well as two-step surface reactions [90,91].

2.5.1 Simple Surface Electrode Reaction

The simplest form of a surface electrode reaction is given by the following Eq. [76–83]:

$$R_{(ads)} \rightleftarrows O_{(ads)} + n e^{-} \tag{2.87}$$

For the sake of simplicity, the charges of the species are omitted. The subscript (ads) implies immobilization of the species on the electrode surface by adsorption, although the adsorption is not the only means of immobilization. The redox species are attributed with their surface concentration Γ, which is a function of time. At the beginning of the experiment, only the R form is present at the electrode surface.

Hence:

$$t = 0: \quad \Gamma_R = \Gamma^*, \quad \Gamma_O = 0 \tag{2.88}$$

In the course of the voltammetric experiment, the total mass of the electroactive material is preserved, which is mathematically expressed by the condition:

$$t > 0: \quad \Gamma_R + \Gamma_O = \Gamma^* \tag{2.89}$$

The derivation of the latter equation with respect to t, yields:

$$\frac{d\Gamma_R}{dt} = -\frac{d\Gamma_O}{dt} \tag{2.90}$$

which means that variations of the surface concentrations of the redox species are equal, but opposite in sign. As the changes of the surface concentrations of the electroactive species prompt current, the following simple differential equations are obtained:

$$\frac{d\Gamma_R}{dt} = -\frac{I}{nFA} \tag{2.91}$$

$$\frac{d\Gamma_O}{dt} = \frac{I}{nFA} \tag{2.92}$$

The solutions of the above differential equations can be readily obtained by integration over the time of the voltammetric experiment, yielding the following solutions:

$$\Gamma_R = \Gamma^* - \int_0^t \frac{I(\tau)}{nFA} d\tau \tag{2.93}$$

$$\Gamma_O = \int_0^t \frac{I(\tau)}{nFA} d\tau \tag{2.94}$$

Considering kinetically controlled process at the electrode surface without lateral interactions between immobilized species, the following form of the Butler–Volmer equation holds:

$$\frac{I}{nFA} = k_{sur} e^{\alpha_a \varphi} \left[\Gamma_R - e^{-\varphi} \Gamma_O \right] \tag{2.95}$$

Here k_{sur} is the surface standard rate constant in units of s^{-1}. By substitution (2.93) and (2.94) into (2.95), one obtains an integral equation, which is a general solution for a surface electrode reaction:

$$\frac{I}{nFA} = k_{sur} e^{\alpha_a \varphi} \left[\Gamma^* - (1 + e^{-\varphi}) \int_0^t \frac{I(\tau)}{nFA} d\tau \right] \tag{2.96}$$

The numerical solution of the above equation is:

$$\Psi_m = \frac{\omega e^{\alpha_a \varphi_m} \left(1 - \frac{1 + e^{-\varphi_m}}{50} \sum_{j=1}^{m-1} \Psi_j \right)}{1 + \frac{\omega e^{\alpha_a \varphi_m}}{50} (1 + e^{-\varphi_m})} \tag{2.97}$$

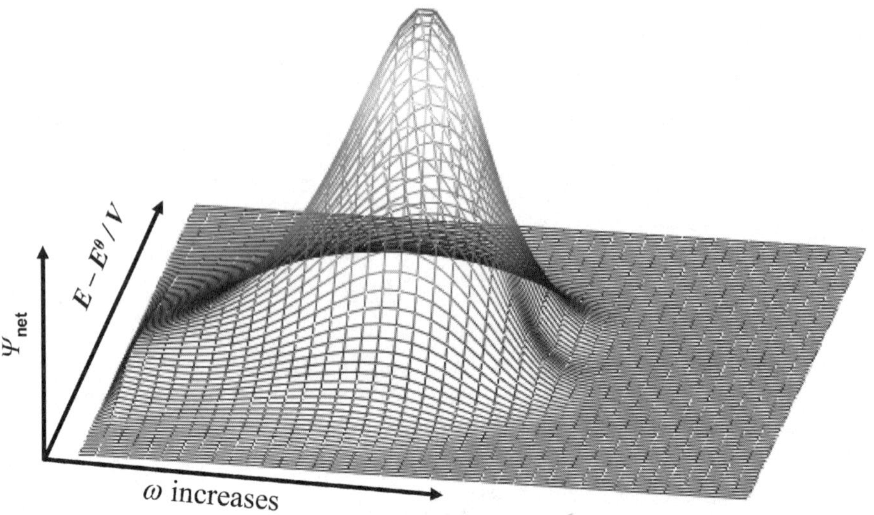

Fig. 2.41 Theoretical net voltammograms simulated for different values of ω. The electrode kinetic parameter increases from *left* to *right* from $\log(\omega) = -1.7$ to 2.2 with an increment of 0.1. Other conditions of the simulations are: $\alpha_a = 0.5$, $nE_{sw} = 50$ mV, $\Delta E = 5$ mV

where the dimensionless current is $\Psi = \frac{I}{nFA\Gamma^* f}$ and $\omega = \frac{k_{sur}}{f}$ is dimensionless electrode kinetic parameter.

The voltammetric profile of the surface reaction depends on the dimensionless kinetic parameter ω, the electron-transfer coefficient α_a, and the amplitude of the potential modulation. The electrochemical reversibility of the electrode reaction is predominantly controlled by the dimensionless kinetic parameter, $\omega = \frac{k_{sur}}{f}$, which unifies the standard rate constant of the electrode reaction and the time window of the voltammetric experiment. Depending on ω voltammetric profiles differ strongly in their shape, magnitude and position. The effect of ω on the net voltammograms is illustrated in Fig. 2.41, whereas Fig. 2.42 depicts all components of the response. Dramatic variations in the net peak magnitude and shape are obvious. If the electrode reaction is either very slow ($\log(\omega) < -1.5$), or very fast ($\log(\omega) > 1.5$), the net response is strongly diminished. Moreover, the net peak starts to split for $\log(\omega) > 0.5$ (see Fig. 2.41), and the net response of a fast electrode reaction consists of two small peaks (Fig. 2.42a). If the electrode reaction behaves as a true reversible process the response totally vanishes. For moderate electrochemical reversibility of the reaction, the magnitude of the response is dramatically enhanced (see Figs. 2.41 and 2.42b).

Figure 2.41 indicates that the net peak current is a parabolic function of the electrode kinetic parameter. This is illustrated in Fig. 2.43. With respect to the electrochemical reversibility of the electrode reaction, approximately three distinct regions can be identified. The reaction is totally irreversible for $\log(\omega) < -2$ and reversible for $\log(\omega) > 2$. Within this interval, the reaction is quasireversible. The parabolic dependence of the net peak current on the logarithm of the kinetic parameter asso-

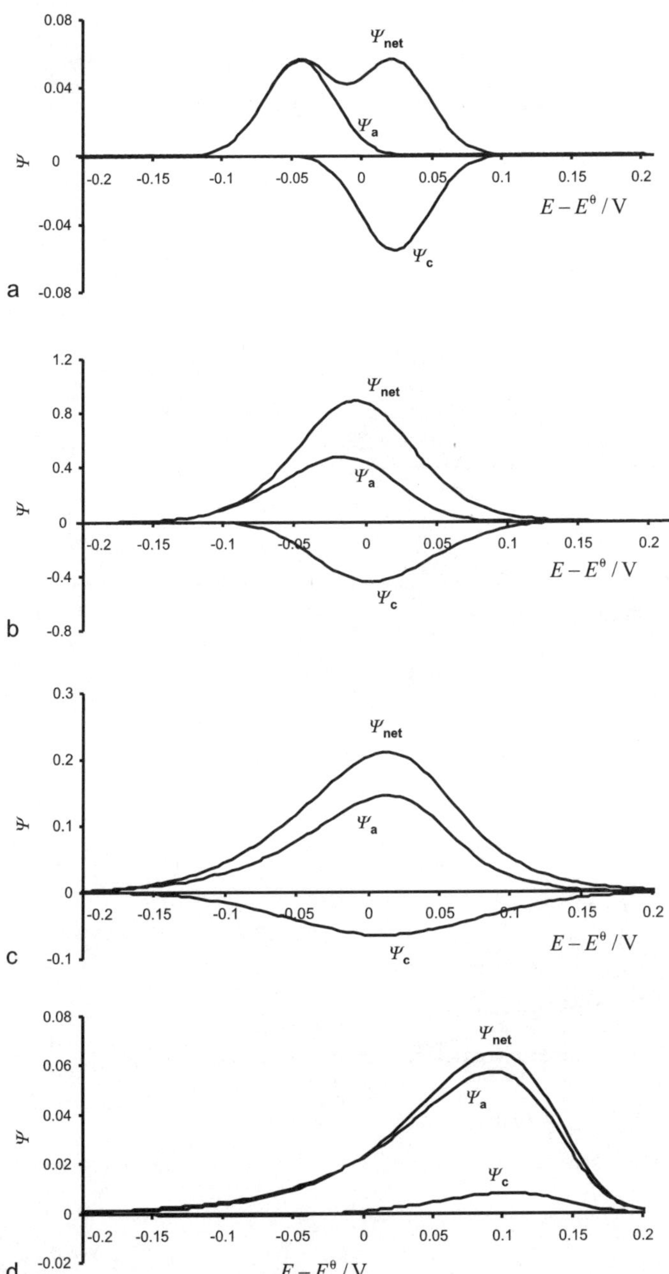

Fig. 2.42a–d Theoretical voltammograms simulated for $\omega = 5$ (**a**); 0.9 (**b**); 0.1 (**c**) and 0.01 (**d**). The other conditions of the simulations are: $\alpha_a = 0.5$, $nE_{sw} = 50$ mV, $\Delta E = 5$ mV. Symbols Ψ_c, Ψ_a, and Ψ_{net} correspond to the cathodic, anodic and net current components of the SW response

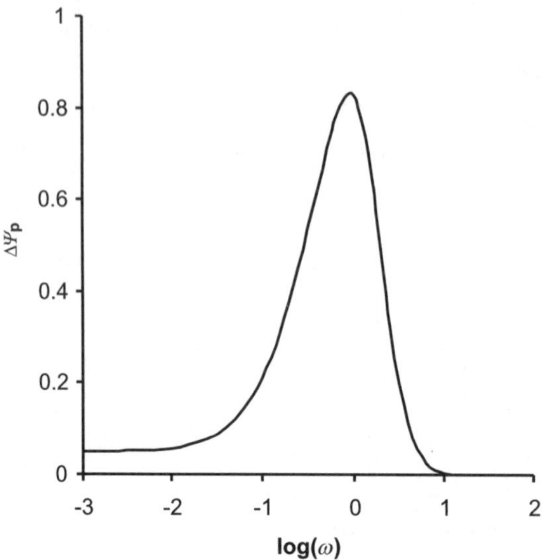

Fig. 2.43 Quasireversible maximum. $\Delta\Psi_p$ as a function of $\log(\omega)$. Conditions of the simulations are the same as for Fig. 2.41

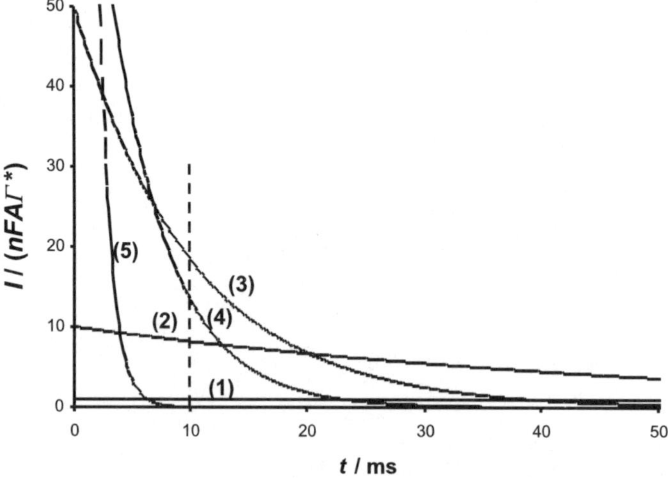

Fig. 2.44 Theoretical chronoamperograms simulated for $E = E^\theta$, $k_{sur} = 1$ (*1*); 10 (*2*); 50 (*3*); 100 (*4*) and 500 s^{-1} (*5*)

ciated with a sharp maximum positioned within the quasireversible region is called "*quasireversible maximum*" [79–82]. A quasireversible maximum is an intrinsic feature of the complete class of surface processes as well as electrode processes coupled to adsorption (Sect. 2.6.1), electrode processes of insoluble salts (Sect. 2.6.3), and electrode processes occurring in a restricted diffusion space (Sect. 2.7).

The quasireversible maximum is a consequence of: (*i*) the current sampling procedure used in SWV, (*ii*) the pulse protocol of SWV, and (*iii*) specific chronoamperometric properties of the surface electrode reaction. A partial explanation of this phenomenon can be provided by analyzing the chronoamperograms of the surface reaction at the formal potential of the reaction. Figure 2.44 shows chronoamperograms simulated for different standard rate constants. Let us assume that the duration of a potential pulse is 10 ms, designated by the vertical line in the graph, which corresponds to the frequency of 50 Hz. If the reaction is very slow (curve 1 in Fig. 2.44), a low current is produced during the potential pulse. If the reaction is fast (curve 5 in Fig. 2.44), a large current is produced at the very beginning of the pulse; however, the current severely diminishes during the potential pulse. As a consequence, at the end of the pulse a minute current remains to be sampled. This is a consequence of the fact that the redox equilibrium between the immobilized redox species is quickly established on the time scale of the experiment. Recall that this phenomenon is absent in the case of a dissolved redox couple (Sect. 2.1.2), since the redox equilibrium is disturbed by the loss of the electroactive material by diffusion. Figure 2.44 shows that the largest currents are measured at the end of the pulse for electrode reactions with a moderate rate (curves 3 and 4 in Fig. 2.44). Interestingly, the maximum response is measured for an electrode reaction attributed with $k_{sur} = 50\,\mathrm{s}^{-1}$, which is equal to the assumed SW frequency. The critical relationship between the frequency and the standard rate constant corresponding to the maximal response can be also revealed from the diagram in Fig. 2.43. The maximum is positioned at $\omega_{max} \sim 1$, which means that the maximum is achieved when the condition $k_{sur} \sim f$ is fulfilled. The frequency satisfying this condition is termed as the "critical frequency", f_{max}. Hence, for a single surface electrode reaction, there is a certain critical frequency that produces the larges dimensionless net peak current. Under such conditions, at potentials close to the formal potential of the redox couple, the rate of the reaction is synchronized with the duration of the pulses enabling repetitive reuse of the immobilized electroactive material. This causes both forward and reverse components of the response to be strongly enhanced, with a minimal separation on the potential axes, producing the highest and the thinnest net peak (see Fig. 2.42b). Based on this inherent property of the surface electrode reaction, a simple methodology for kinetic measurements has been developed [82]. It requires only the determination of the critical value of the kinetic parameter ω_{max} by means of simulations, and determination of the critical frequency f_{max} in the experiment. Accordingly, the standard rate constant can be calculated through simple relation:

$$k_{sur} = \omega_{max} f_{max} \tag{2.98}$$

The exact values of the critical kinetic parameter depend on the electron-transfer coefficient and the amplitude. These values are listed in Table 2.3. If the electron-transfer coefficient is not known, an average value of the critical kinetic parameter $(\omega_{max})_{avr}$ can be used. The values of $(\omega_{max})_{avr}$ for different amplitudes are given in Table 2.4. The error in the estimation of k_{sur} by using ω_{max} is close to 10%.

Beside the quasireversible maximum, the splitting of the net peak is the second intriguing feature of a surface electrode reaction [84]. The splitting emerges by

Table 2.3 Dependence of the critical kinetic parameter ω_{max} on the normalized amplitude nE_{sw} and electron-transfer coefficient α_a. Conditions of the simulations are the same as for Fig. 2.41

α_a	ω_{max} nE_{sw}/mV				
	15	25	30	40	50
0.9	1.43	1.35	1.38	1.33	1.26
0.8	1.32	1.3	1.25	1.17	1.08
0.7	1.31	1.26	1.2	1.1	0.97
0.6	1.29	1.2	1.16	1.04	0.9
0.5	1.28	1.19	1.13	1.01	0.88
0.4	1.27	1.18	1.13	1.02	0.89
0.3	1.26	1.22	1.17	1.04	0.94
0.2	1.25	1.24	1.19	1.12	1.04
0.1	1.25	1.27	1.3	1.26	1.19

increasing ω and the amplitude of the potential modulation. The intrinsic reasons causing splitting are rather complex and they originate from the potential dependence of the forward and reverse rate constants of the electrode reaction and specific chronoamperimetric features of the surface reaction. Having no intention to provide a rigorous mathematical prove for the splitting, the following discussion is focused on the qualitative description of this phenomenon and its utilization for kinetic and thermodynamic characterization of the surface reaction. Figure 2.45 depicts the variation of the shape of the response under influence of the kinetic parameter. The peak potential separation between the forward and reverse component increases in proportion to ω, which causes broadening of the net peak width (Fig. 2.45b), ending with a splitting of the net peak (Fig. 2.45c). It is interesting to note that by increasing of ω the forward (oxidation) and the reverse (reduction) components shift toward more negative and more positive potentials, respectively. This is completely opposite compared to the corresponding situation in cyclic voltammetry, or compared to the quasireversible reaction of a dissolved redox couple under conditions of SWV. Hence, in case of the split net peaks, the more positive peak reflects the reduction process, whereas the more negative peak represents the oxidation process. Therefore, the splitting of the net peak, together with the inverse position of the oxidation and reduction currents, provide a clear criterion for a qualitative recognition

Table 2.4 Dependence of the average critical kinetic parameter $(\omega_{max})_{avr}$ on the normalized amplitude nE_{sw} valid for the electron-transfer coefficient $0.1 < \alpha_a < 0.9$. Conditions of the simulations as in Fig. 2.41

nE_{sw}/mV	$(\omega_{max})_{avr}$
15	1.30 ± 0.05
25	1.25 ± 0.07
30	1.21 ± 0.08
40	1.12 ± 0.11
50	1.02 ± 0.14

and distinguishing a surface reaction from all other diffusion controlled electrode processes. As illustrated by Fig. 2.46, the splitting is strongly sensitive to the normalized SW amplitude, nE_{sw}. For a given kinetic parameter and number of electrons, there is a minimal amplitude causing splitting. These critical values of the SW amplitude can serve for a rough estimation of the electrode kinetic parameter in a particular experiment (Table 2.5). As shown in the inset of Fig. 2.46, the potential separation between split peaks ΔE_p varies linearly with nE_{sw}. All the lines in the inset have the same slope, only the intercept depends on ω. The pronounced sensitivity of the split peaks to the SW amplitude reveals that the latter instrumental parameter is a second tool, besides the frequency, for inspection of the kinetics of the electrode reaction.

In addition, the split peaks can be used for estimation of electron-transfer coefficient as well as for precise determination of the formal potential of the surface electrode reaction. The potential separation between split peaks is insensitive to the electron-transfer coefficient. However, the relative ratio of the heights of the split peaks depends on the electron-transfer coefficient according to the following function:

$$\frac{\Psi_{p,c}}{\Psi_{p,a}} = 5.64 \exp(-3.46\alpha_a) \qquad (2.99)$$

With respect to the formal potential of the surface electrode reaction, Figs. 2.45c and 2.46 show that the split peaks are symmetrically located around the formal potential, which enables precise determination of this important thermodynamic parameter.

Numerous experimental systems verified the theory of surface electrode reactions. Reductions of methylene blue [92], azobenzene [79, 82] alizarine red S [93], probucol [94], molybdenum(V)-fulvic acid complex [95], molybdenum(VI)-1,10 phenanthroline-fulvic acid complex [96], indigo [97], and reduction of vanadium(V) [98] at a mercury electrode are some of the examples for surface electrode

Table 2.5 Critical values of the SW amplitude and corresponding potential separations of the split peaks for various values of the electrode kinetic parameter. Conditions of the simulations are the same as for Fig. 2.41

ω	$(nE_{sw})_{min}/mV$	$\Delta E_p/mV$
0.8	100	80
1	90	80
2	60	55
3	50	60
4	50	80
5	50	80
6	50	80
7	40	70
8	40	50
9	40	70
10	40	75

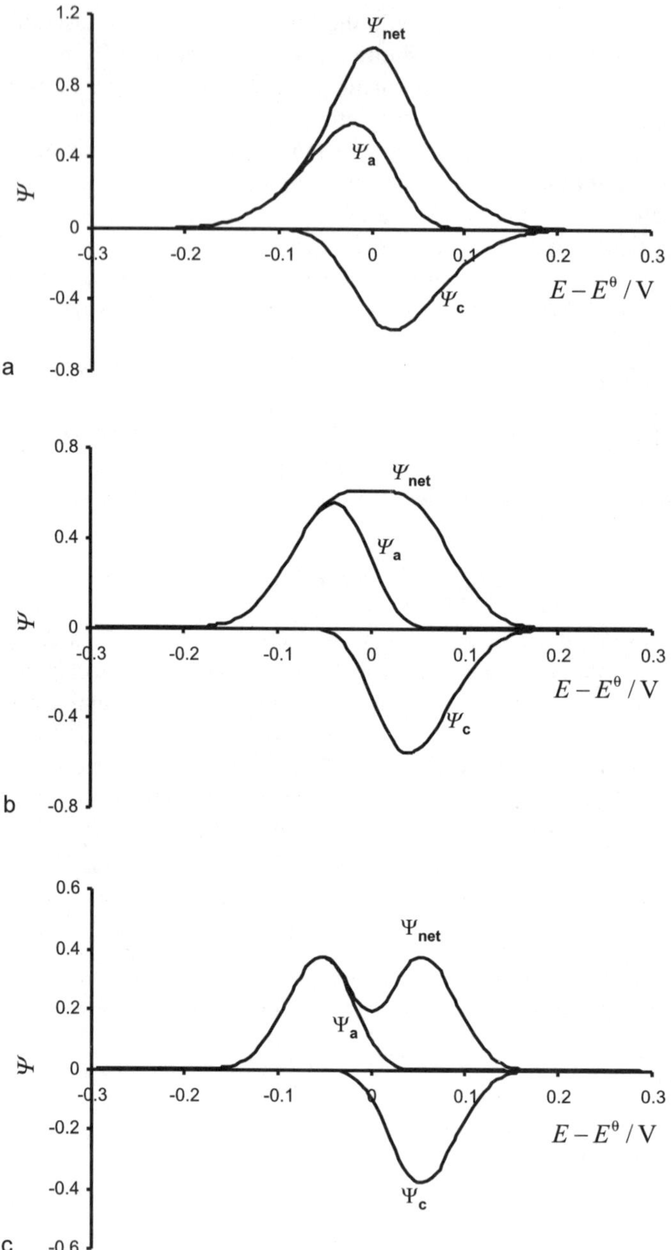

Fig. 2.45a–c Theoretical voltammograms simulated for $\omega = 0.5$ (**a**); 1.2 (**b**) and 2 (**c**). The other conditions of the simulations are: $\alpha_a = 0.5$, $nE_{sw} = 80$ mV, and $\Delta E = 5$ mV

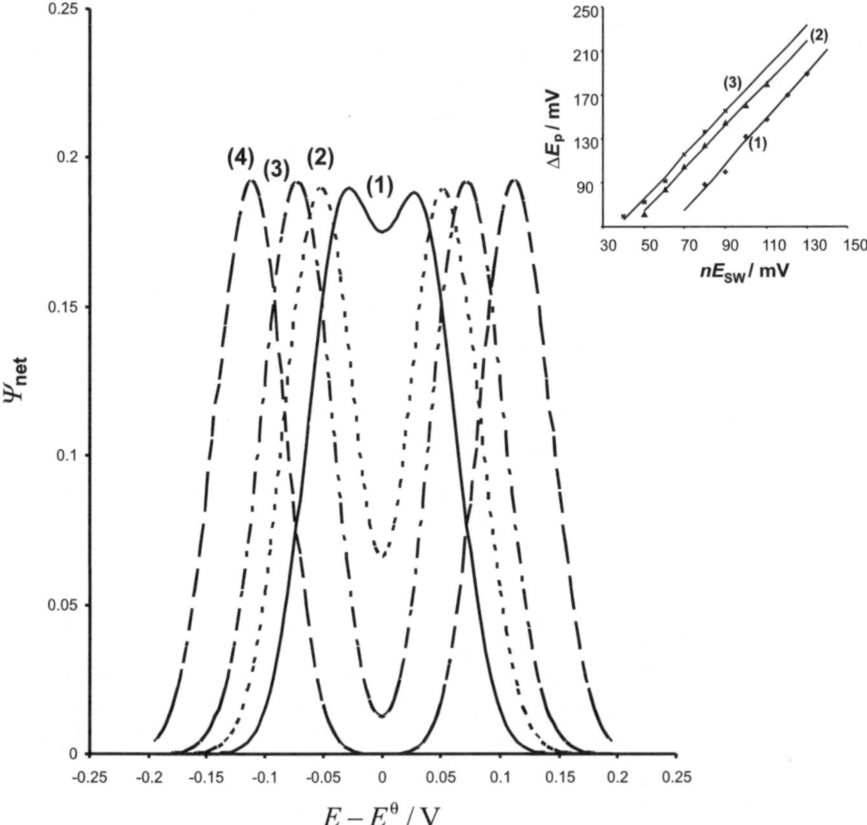

Fig. 2.46 Theoretical net voltammograms simulated for $nE_{sw} = 50$ (*1*); 70 (*2*); 90 (*3*) and 130 mV (*4*). The other conditions of the simulations are: $\log(\omega) = 0.5$, $\alpha_a = 0.5$, and $\Delta E = 5$ mV. The *inset* shows the dependence of the peak potential separation ΔE_p on the product nE_{sw} for $\omega = 1.5$ (*1*); 3 (*2*) and 5 (*3*)

reactions studied by SWV. Another set of studies refer to important biochemical systems such as cytochrome *c* immobilized on carboxylic acid-terminated self-assembled monolayer of alkenthiol on the gold electrodes [76], myoglobin incorporated into thin films of didodecyldimethylammonium bromide immobilized on a basal plane pyrolytic graphite electrode [78], myoglobin immobilized on the titanate nanotubes [99] flavocytochrome (the globular fumarate reductase and the membrane-bound fumarate reductase) deposited at the edge plane pyrolytic graphite electrode [100], azurin (a blue copper protein) adsorbed on the edge plane pyrolytic graphite [101] and paraffin-impregnated graphite electrode [83] or immobilized with the aid of self-assembled monolayers of various 1-alkenthiols on a gold electrode [101].

Reeves et al. [76] suggested first the methodology to measure the rate constant of the surface reaction of cytochrome *c* on the basis of potential separation between the forward and reverse components of the SW response. O'Dea et al. [77] analyzed the reduction of adsorbed azobenzene at a mercury electrode using the curve fitting method based on COOL algorithm. The experimental voltammograms of azobenzene fit well with the theoretical predictions exhibiting both the typical properties of the surface reaction such as quasireversible maximum and split SW peaks.

Using the quasireversible maximum, together with the splitting of the net peak, the charge transfer kinetics of azobenzene, [82], alizarine red S [84, 93], probucol [94], molybdenum(VI)-1,10 phenanthroline-fulvic acid complex [96], indigo [97], vanadium(V) [98], cercosporin [102] and altertoxin I [103], both mycotoxins (fungal metabolites), have been studied. The strategy based on quasireversible maximum and split SW peaks is both simple and versatile. For quasireversible maximum, one requires the inspection of the dependence of ratio between the real net peak current and corresponding frequency ($\frac{\Delta I_p}{f}$) vs. the frequency. Note that the ratio $\frac{\Delta I_p}{f}$ corresponds to the dimensionless peak current $\Delta \Psi_p$. Hence, the dependence $\frac{\Delta I_p}{f}$ vs. $\log(f)$ is a parabolic function, resembling the theoretical dependence presented in Fig. 2.43, with a maximum located at critical frequency f_{max}, that satisfies the condition

$$f_{max} = \frac{k_{sur}}{\omega_{max}} \tag{2.100}$$

Knowing the critical values of ω_{max}, given in Table 2.3, the surface standard rate constant is estimated using (2.98). Experimental quasireversible maxima of azobenzene measured at various pH are shown in Fig. 2.47 [79]. The important advantage of the quasireversible maximum is its slight sensitivity to the electron-transfer coefficient and the number of electrons exchanged, and insensitivity to the surface concentration of the electroactive material. Without having an initial knowledge of α_a and n, k_{sur} can be estimated with an error of about 10% on the basis of the quasireversible maximum.

The splitting of the net peak can be used for kinetic measurements by comparing the potential separation between the split peaks of theoretical and experimental voltammograms. As the splitting appears when the ω is sufficiently large, one decreases the frequency to achieve splitting of the experimental voltammograms. In general, the lower the frequency, the higher the potential separation between the split peaks. For a given low frequency, the splitting is analyzed by varying the SW amplitude. It is important to note that the splitting is strongly sensitive to nE_{sw}. Hence, contrary to the quasireversible maximum, the splitting is highly sensitive to the number of electrons involved in the electrode reaction. Figure 2.48 shows the splitting of the net voltammograms of alizarine red-S adsorbed at the mercury electrode under influence of the signal amplitude. The strategy based on splitting is very useful for kinetic measurements of a very fast surface electrode reaction. Although the reaction is very fast, the kinetic measurements are performed at low frequency, i.e., with a long duration of potential pulses.

Marchiando et al. employed both the quasireversible maximum and splitting of the net SW peak for a complete kinetic and thermodynamic characterization of alter-

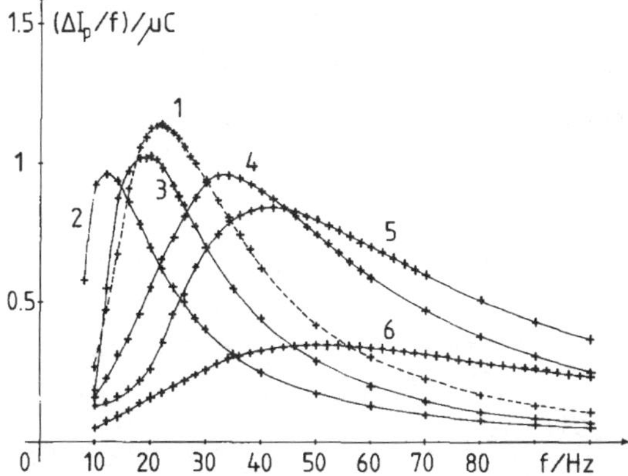

Fig. 2.47 Quasireversible maxima measured with reduction of 1.5×10^{-5} (1) and 1×10^{-5} mol/L (2–6) azobenzene solution at a mercury electrode. Supporting electrolytes: 0.25 mol/L acetate buffer (pH 4.65, *curve 1*), 0.5 mol/L NaClO$_4$ buffered to pH 4 (2); 3 (3); 2 (4) and 1 (5), and 1 mol/L HClO$_4$ (6) (reprinted from [79] with permission)

toxin I [103] and cercosporin [102], immobilized on glassy carbon electrodes. The quasireversible maxima, measured at various bulk concentrations of cercosporin, are given in Fig. 2.49. The average critical frequency is $f_{\max} = (282 \pm 38)\,\mathrm{Hz}$. As the number of electrons n and the electron-transfer coefficient α_c are not know, an average value of the critical kinetic parameter $(\omega_{\max})_{\mathrm{avr}}$ from Table 2.4 can be used. Assuming $n = 1$, the average value of the kinetic parameter is $(\omega_{\max})_{\mathrm{avr}} = 1.25 \pm 0.07$ for $E_{\mathrm{sw}} = 25$ mV. If $n = 2$, the value of $(\omega_{\max})_{\mathrm{avr}} = 1.02 \pm 0.14$ should be used, for the same amplitude. Thus, the standard rate constant k_{sur} was estimate to range within the interval from $(3.5 \pm 0.7) \times 10^2\,\mathrm{s}^{-1}$ to $(2.9 \pm 0.8) \times 10^2\,\mathrm{s}^{-1}$. The number of electrons can be estimated by analyzing the splitting as a function of the amplitude, at a constant frequency of the potential modulation. Figure 2.50 shows representative voltammograms of cercosporin recorded at amplitudes of 25, 75, and 125 mV. As previously mentioned, the slope of the dependence ΔE_{p} vs. E_{sw} is particularly sensitive to the number of electrons. Accordingly, comparing the slope of the experimental to theoretical dependencies, one can estimate the number of exchanged electrons. For the measurements with cercosporin conducted at $f = 5, 10, 15, 30, 40, 50$ and 200 Hz, the average slope of the dependence ΔE_{p} vs. E_{sw} is 1.48 ± 0.14. Simulations conducted under corresponding conditions, taking k_{sur} within the interval from 290–350 s^{-1}, revealed that the number of electrons is one. Bearing in mind that the split peaks are symmetrically located around the formal potential of the surface electrode reaction, the formal potential of cercosporin at a glassy carbon electrode was estimated to be $E_c^{\theta\prime} = (-0.260 \pm 0.011)\,\mathrm{V}$ vs. SCE. Furthermore, the average value of the ratio $I_{\mathrm{p,a}}/I_{\mathrm{p,c}}$ was 1.0 ± 0.1; hence the cathodic electron-transfer coefficient estimated from (2.99) is 0.50 ± 0.03. Having in hand the values for n and α_c, one can re-evaluate k_{sur} from the critical frequency measured at the quasireversible maximum, selecting the exact value for the critical kinetic parameter

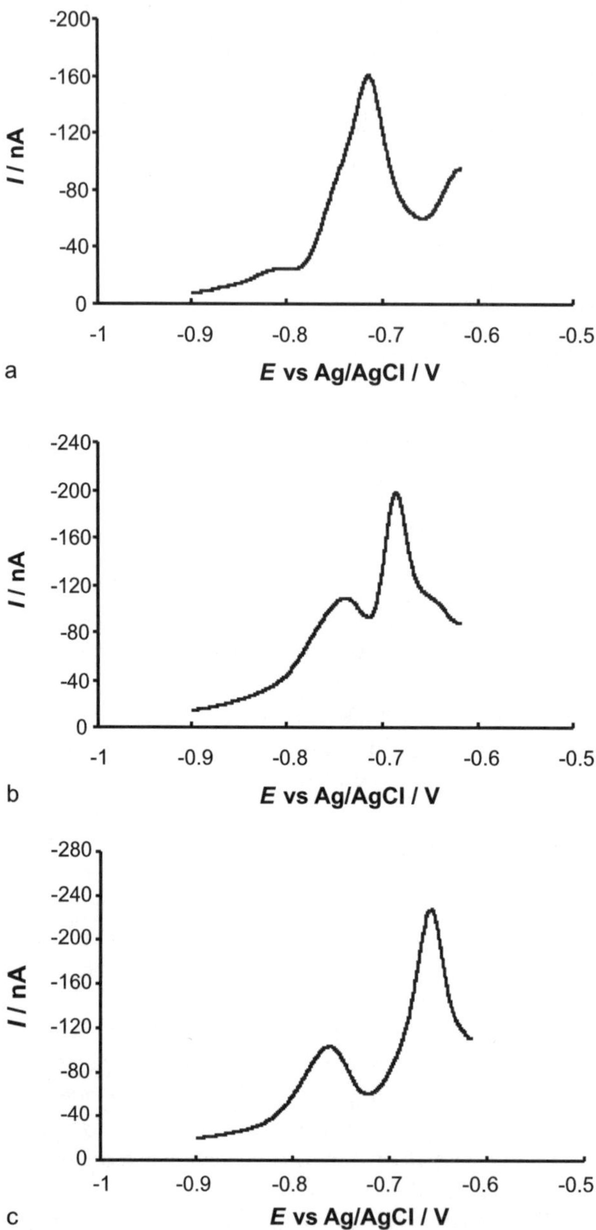

Fig. 2.48a–c Net voltammograms for the reduction of 1×10^{-6} mol/L alizarine-red S solution in 1 mol/L KNO_3 buffered with a borate buffer to pH 9. The SW amplitude is 30 (**a**); 60 (**b**) and 90 mV (**c**). The other conditions are: $t_{acc} = 5$ s, $E_{acc} = -0.1$ V vs. Ag/AgCl (3 mol/L KCl), $f = 30$ Hz, and $\Delta E = 0.1$ mV (a courtesy of Dr. François Quentel from the Laboratoire de Chimie Analytique, UMR-CNRS 6521, Brest, Université de Bretagne Occidentale, France)

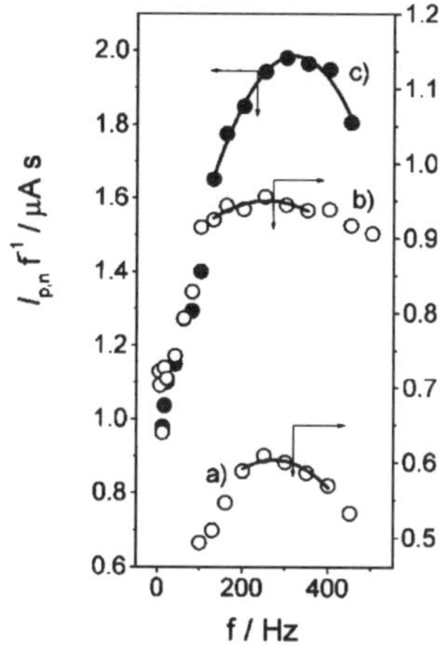

Fig. 2.49 Quasireversible maxima of cercosporin measured in $1\,\text{mol/L}$ $HClO_4$ aqueous solution at a glassy carbon electrode. The electrode is modified in a separate solution containing $3.74\,\mu\text{mol/L}$ (**a**), $5.01\,\mu\text{mol/L}$ (**b**), and $18.7\,\mu\text{mol/L}$ (**c**) cercosporin. Other experimental conditions are: $E_{\text{acc}} = -0.7\,\text{V}$ vs. SCE, $t_{\text{acc}} = 30\,\text{min}$, $E_{\text{sw}} = 25\,\text{mV}$, $\Delta E = 5\,\text{mV}$ (reprinted from [102] with permission)

$\omega_{\max}$ from Table 2.3. For $nE_{\text{sw}} = 25\,\text{mV}$, and $\alpha_{\text{c}} = 0.5$, the critical kinetic parameter is $\omega_{\max} = 1.19$. Thus, the standard rate constant of cercosporin is $k_{\text{sur}} = (336 \pm 46)\,\text{s}^{-1}$.

Important experimental examples for surface electrode reactions are those concerning redox enzymes and proteins [76, 78, 83, 99–101]. The overall methodology based on immobilization of proteins on the surface of a solid electrode in order to interrogate the redox properties of a protein with the aid of voltammetric techniques is termed protein film voltammetry [100, 101]. To account for intricate voltammetric features of these complex molecules under large overpotentials, in the modeling of a surface electrode mechanism Butler–Volmer kinetics has been frequently replaced by Marcus kinetic theory [78, 101]. According to the latter theory, the electron-transfer rate to and from an electrode increases exponentially with modest electrochemical driving force and then reaches saturation level to a maximum value at sufficiently large overpotential. Recall that according to the simpler Butler–Volmer model, the electrochemical rate constants increases exponentially with the overpotential without reaching any limiting value. Combining the Markus equation for non-adiabatic electron transfer with Fermi–Dirac distribution for electronic states in the electrode results in the following equation for the constants corresponding to the oxidation (k_{ox}) and reduction (k_{red}) processes [101]:

$$k_{\text{ox/red}} = k_{\max}\sqrt{\frac{RT}{4\pi\lambda}} \int_{-\infty}^{\infty} \frac{\exp\left(\frac{\lambda \pm F(E-E_{\text{c}}^{\theta'})}{RT} - x\right)^2 \frac{RT}{4\lambda}}{\exp(x) + 1}\,dx \tag{2.101}$$

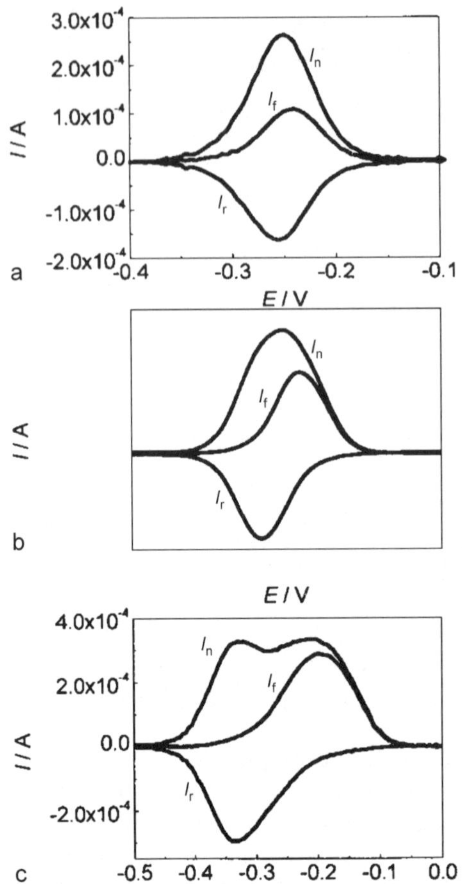

Fig. 2.50a–c The forward (I_f), reverse (I_r) and the net (I_{net}) components of the SW response of cercosporin recorded at $E_{sw} = 25$ (**a**), 75 (**b**), and 125 mV (**c**). The other experimental conditions are: $f = 200$ Hz, $\Delta E = 5$ mV, c(cercosporin)=1.87 μmol/L. The other conditions are the same as for Fig. 2.49 (reprinted from [102] with permission)

Here, $E_c^{\theta'}$ is the formal potential of the immobilized redox couple, λ (in $\mathrm{J\,mol^{-1}}$) is the reorganization energy required for the heterogeneous electron-transfer reaction, and k_{max} is the maximum rate constant at sufficiently large overpotentials. Knowing k_{max} and λ, one can calculate the standard exchange rate constant k_0 at zero driving force ($E = E_c^{\theta'}$), which corresponds to the standard rate constant k_{sur} in the Butler–Volmer model. The latter kinetic model can provide satisfactory results in modeling of the processes in protein film voltammetry if the applied overpotentials are significantly smaller than the reorganization energy λ. On the other hand, when the SW voltammetric experiment in protein film voltammetry is conducted at large amplitudes, i.e., large overpotentials, one can estimate the reorganization energy as well as the limiting value of the rate constant k_{max}.

Rusling et al. [78] first modeled a surface electrode reaction under conditions of SWV on the basis of Markus kinetic theory, in order to study myoglobin, an oxygen-transport protein, incorporated into thin films of didodecyldimethylammonium bromide on the surface of a basal plane pyrolytic graphite electrode. Figure 2.51 shows the forward and backward components of the SW response of myoglobin recorded at large amplitudes, where points are experimental data and lines are best fits by non-linear regression analysis onto the Marcus model. Using the nonlinear regression analysis, the electrochemical rate constants together with the reorganization energies have been estimated from single experiments measured. The heterogeneous rate constants are $\log(k_{\mathrm{red}}^{\mathrm{null}}) = (3.3 \pm 0.8)$ s^{-1} and $\log(k_{\mathrm{ox}}^{\mathrm{null}}) = (3.3 \pm 0.8)$ s^{-1}. Here, k^{null} is the heterogeneous rate constant in units of s^{-1} when the overpotential η is equal to the reorganization energy λ, i.e., $\lambda + \eta = 0$, where $\eta = E - E^{\theta'}$. The mean value for reorganization energy λ_{red} was 0.41 ± 0.02 eV and $n\lambda_{\mathrm{ox}}$ was 0.21 ± 0.01 eV.

Armstrong et al. [101] modeled the surface electrode reaction of azurin, a blue copper protein, adsorbed on edge plane pyrolytic graphite and gold electrodes modified with different self-assembled monolayers of various 1-alkenthiols, on the basis of Markus theory. These authors emphasized the utility of splitting of the SW response under large amplitudes for estimation of the reorganization energy and maximum rate constant. Figures 2.52 shows the effect of the SW frequency on the response of azurin adsorbed on a edge plane pyrolytic graphite electrode, revealing clearly the splitting of the net SW peak. Furthermore, Fig. 2.53 shows the splitting by varying the SW frequency for the experiment with azurin adsorbed on a 1-decanethiol-modified gold electrode. These authors concluded that the standard electron exchange rate constant k_0 is dependent on the electrode type used, whereas

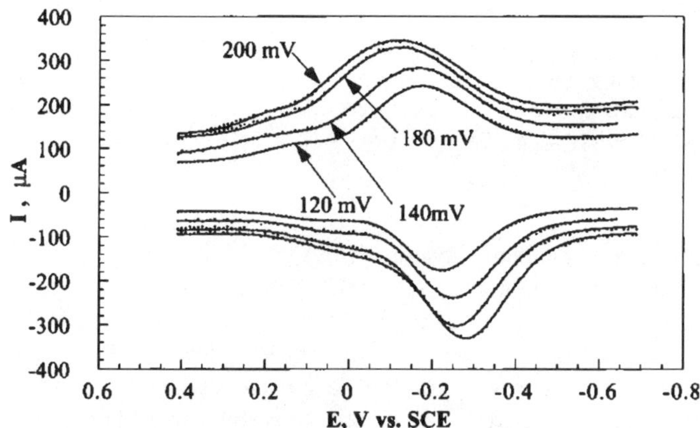

Fig. 2.51 Forward and reverse square-wave voltammograms of myoglobin-didodecyldimethyl-ammonium bromide films on a basal plane pyrolytic graphite electrodes at 200 Hz frequency, 10 mV step height, and different pulse heights. *Points* are experimental data, and *lines* are best fits by nonlinear regression onto the Marcus model. Background currents are included in experimental and computed data. $T = 37.0 \pm 0.2$ °C, and the supporting electrolyte is 20 mmol/L pH 6.0 phthalate buffer +180 mmol/L NaCl (reprinted from [78] with permission)

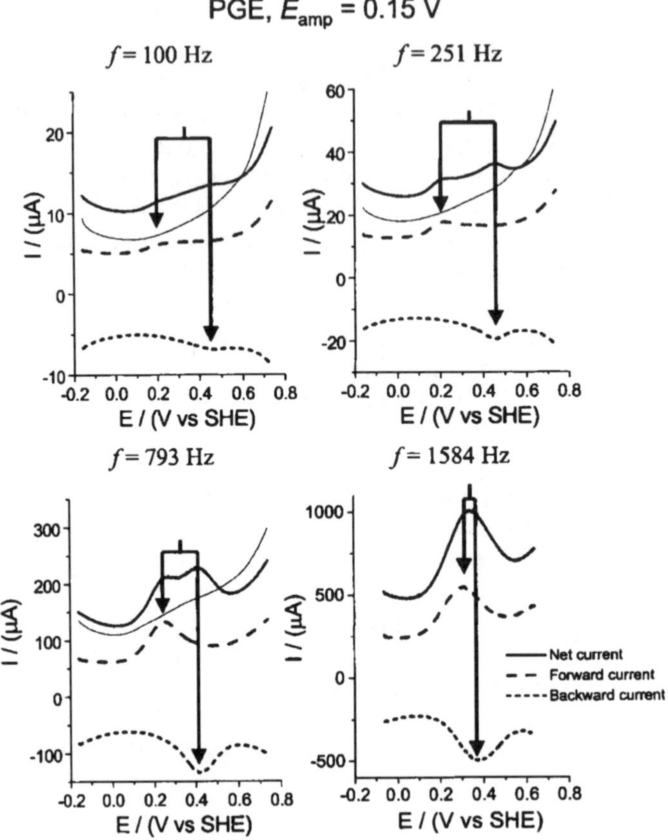

Fig. 2.52 "Raw" square-wave voltammetry data of azurin immobilized on edge plane pyrolytic graphite electrode (PGE) (pH 4.0, 2.0 mol/L NaCl) at different frequencies with $E_{sw} = 0.15$ V. *Dashed line*: forward current. *Dotted line*: backward current. *Solid thick line*: net current; *solid thin line*: blank net current (reprinted from [101] with permission)

the maximum rate constant is essentially invariant with a rate of $(6 \pm 3) \times 10^3 \, \text{s}^{-1}$, at $0\,^{\circ}\text{C}$ for both oxidation and reduction, irrespective of whether the graphite or gold-modified electrode is used. Interestingly, using the Markus theory, the fitting of the experimental and theoretical results yielded an extremely low value for the reorganization energy ($\lambda < 0.25\,\text{eV}$); however, good fits have been obtained by using an alternative model in which the electron transfer was gated by a preceding chemical process, involving highly ordered protein configuration on the electrode surface [101].

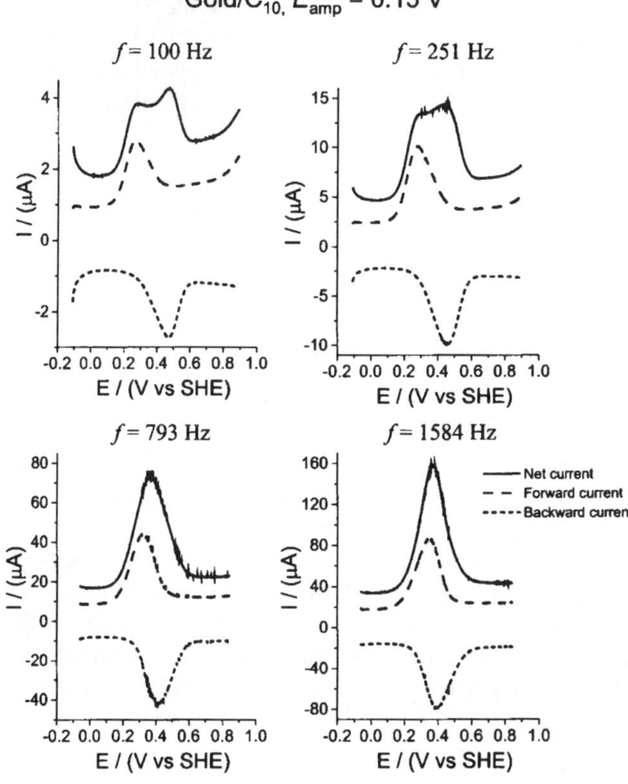

Fig. 2.53 "Raw" square-wave voltammetry data of azurin immobilized on 1-decanethiol-modified gold (pH 4.0, 0.1 mol/L Na_2SO_4) at different frequencies with $E_{sw} = 0.15$ V. *Dashed line*: forward current. *Dotted line*: backward current. *Solid line*: net current (reprinted from [101] with permission)

2.5.2 Surface Electrode Reaction Involving Interactions Between Immobilized Species

The theory described in the previous chapter has been developed under provision that no interactions exists between immobilized species. However, for many experimental systems, this prerequisite is not fulfilled [85, 104, 105]. Hence, it is of interest to consider a case of a surface reaction involving lateral interactions. In a general case, various sorts of interactions can be assumed between O and R forms immobilized on the electrode surface [106]. The following discussion is restricted to the case of uniform interactions between immobilized species.

To describe a surface electrode reaction in the presence of uniform interactions, besides (2.88) to (2.92), the following form of the kinetic equation is required:

$$\frac{I}{nFA} = k_{sur} e^{\alpha_a \varphi} e^{-2a\theta} \left[\Gamma_R - e^{-\varphi} \Gamma_O \right] \qquad (2.102)$$

The intrinsic parameter, characterizing the type of interactions, is the Frumkin interaction parameter a, which is positive for attractive forces and negative for repulsive forces. In addition, $\theta = \frac{\Gamma^*}{\Gamma_{\max}}$ is the fraction of the electrode covered with deposited material, and $\Gamma_{\max}$ is the maximal surface coverage. Combining (2.93) and (2.94) with (2.102), the following integral equation is obtained as a general solution:

$$\frac{I}{nFA} = k_{\mathrm{sur}}\, e^{\alpha_a \varphi}\, e^{-2a\theta} \left[\Gamma^* - \left(1 + e^{-\varphi}\right) \int_0^t \frac{I(\tau)}{nFA}\, d\tau \right] \tag{2.103}$$

The numerical solution of the above equations is:

$$\Psi_m = \frac{\omega\, e^{\alpha_a \varphi_m}\, e^{-2a\theta} \left(1 - \frac{1+e^{-\varphi_m}}{50} \sum\limits_{j=1}^{m-1} \Psi_j \right)}{1 + \frac{\omega\, e^{\alpha_a \varphi_m}\, e^{-2a\theta}}{50} \left(1 + e^{-\varphi_m}\right)} \tag{2.104}$$

In the presence of interactions, besides the kinetic parameter ω and the electron-transfer coefficient, the response is controlled by the interaction product $a\Theta$. The kinetic parameter and the interaction product can be unified into a single complex kinetic parameter defined as $\omega_{\mathrm{int}} = \frac{k_{\mathrm{sur}}}{f}\, e^{-2a\theta}$. Introducing this parameter, the solution (2.104) simplifies to that given by (2.97) valid for a simple surface reaction. Consequently, the effect of ω_{int} is equivalent to the effect of ω, elaborated in the previous chapter. Therefore, the overall effect of interactions is quite predictable.

Studying a single surface reaction without interactions, the kinetic parameter can be varied by altering the frequency of the potential modulation. However, in the

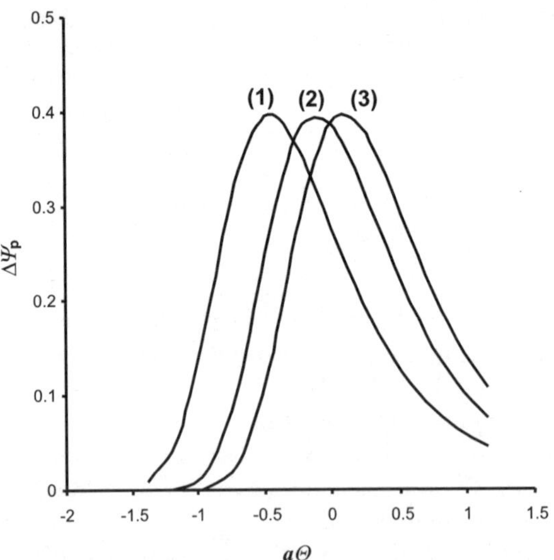

Fig. 2.54 Dimensionless net peak current as a function of the interaction product $a\Theta$ for $(k_{\mathrm{sur}}/f) = 0.5$ (*1*); 1 (*2*) and 1.5 (*3*). The other conditions of the simulations are: $\alpha_a = 0.5$, $nE_{\mathrm{sw}} = 20\,\mathrm{mV}$, and $\Delta E = 10\,\mathrm{mV}$

presence of interactions, ω_{int} can be affected by both the frequency, or the relative surface coverage of the electrode Θ. This is the most important difference between a simple surface reaction and a reaction involving interactions between immobilized species. If the interactions are attractive, $a > 0$, increasing Θ causes a decrease of the parameter ω_{int}, i.e., a decrease of the electrochemical reversibility of the electrode reaction. As a consequence, in an oxidative electrode mechanism, the attractive interactions make the surface reaction thermodynamically unfavorable, causing a shift of the peak potential in positive direction. For repulsive forces the situation is opposite. With any sort of interactions, the net peak current, together with the half-peak width increase or decrease with Θ, depending on the value of the ratio k_{sur}/f. Figure 2.54 shows the net peak current as a function of the interaction product $a\Theta$, over a wide range of values. Basically, these curves are quasireversible maxima constructed by variation of the interaction product. The position of the maximum is associated with a certain critical value of the interaction product $(a\Theta)_{max}$, which depends on the value of the ratio k_{sur}/f. To reach the quasireversible maximum, the following condition must be fulfilled:

$$\frac{k_{sur}}{f} e^{-2(a\theta)_{max}} = (\omega_{int})_{max} \tag{2.105}$$

The values of $(\omega_{int})_{max}$ are identical with ω_{max} for the simple surface electrode reaction given in Table 2.3.

From the definition of ω_{int} follows that the quasireversible maximum can be also determined by varying the frequency, while keeping Θ constant. In analogy to (2.105), it is obvious that the critical frequency, associated with the position of the maximum, depends on the interaction product $a\Theta$. The relationship between the critical frequency and the interaction product is given by the following equation:

$$\ln(f_{max}) = \ln\left[\frac{k_{sur}}{(\omega_{int})_{max}}\right] - 2a\theta \tag{2.106}$$

This equation is of particular importance since it enables estimation of both the interaction product a and the standard rate constant k_{sur}, provided the relative surface coverage is known. For this, the quasireversible maximum is to be determined by varying the frequency for various values of the surface coverage Θ. Plotting $\ln(f)$ vs. Θ, a line is obtained with a slope and intercept equal to $-2a$ and $\ln\left[\frac{k_{sur}}{(\omega_{int})_{max}}\right]$, respectively.

The theory for surface electrode reaction involving interactions has been applied to study the electrode reduction of probucol [104] and Mo(VI) in the presence of phenanthroline and fulvic acids [105] at a mercury electrode. Probucol is a cholesterol-lowering agent, which undergoes one electron reduction at a mercury electrode to form a stable anion radical [107]. Both forms of the redox couple are strongly adsorbed on the mercury electrode surface. The net voltammograms of probucol reduction at a various accumulation time are illustrated in Fig. 2.55. The prolongation of the accumulation from 15 to 200 s causes the response to increase in its height, while its position is shifted slightly towards more positive potentials. However, a further increase of the accumulation time results in a substantial decrease of the peak current together with a strong shift of the response in the positive direction. The dependence of the net peak current on the accumulation time is depicted in Fig. 2.56 (curve 1). The parabolic shape of curves 1 and 2 indicates clearly the presence of significant interactions between the adsorbed molecules. In addition, the shift of the peak in a positive direction shows that the reduction is energetically favored by increasing of the surface coverage, implying repulsive forces within the deposited film. Recall that the accumulation time affects the interaction product $a\Theta$ through the surface coverage Θ. The interaction product can also be varied by changing the Frumkin interaction parameter a. For these purposes, a series of experiments have been carried out in the presence of a certain amount of acetonitrile that affects the solubility of probucol, thus affecting the Frumkin interaction parameter of the adsorbed species. For these reasons, in the presence of 3% (v/v) acetonitrile, the peak currents of probucol are considerably suppressed (curve 2 in Fig. 2.56). The shape of the $\Delta I_p - t_{acc}$ relationship is still parabolic, but the maximum of the parabola is displaced slightly toward the longer accumulation time. In the presence of 6% (v/v) acetonitrile, the interactions between probucol molecules disappear completely and consequently the $\Delta I_p - t_{acc}$ relationship obeys a Langmuir adsorption isotherm law (curve 3 in Fig. 2.56).

Using the aforementioned methodology, the electrode reaction of Mo(VI) has been studied in the presence of phenanthroline and an excess of fulvic acids [105]. Both ligands exhibit a synergetic effect toward adsorption of the mixed complex of

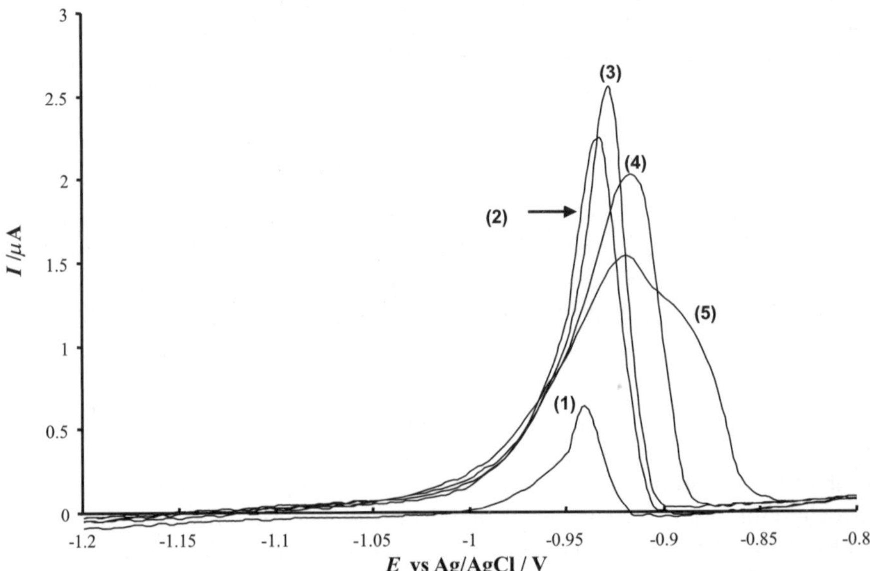

Fig. 2.55 The effect of accumulation time on the net peaks for the reduction of 4×10^{-6} mol/L probucol solution in 1 mol/L KNO$_3$. The experimental conditions are $t_{acc} = 15$ (1), 100 (2), 200 (3), 300 (4) and 400 s (5). The other conditions are: $E_{acc} = -0.7$ V, $f = 150$ Hz, $E_{sw} = 25$ mV, and $\Delta E = 4$ mV (with permission from [85])

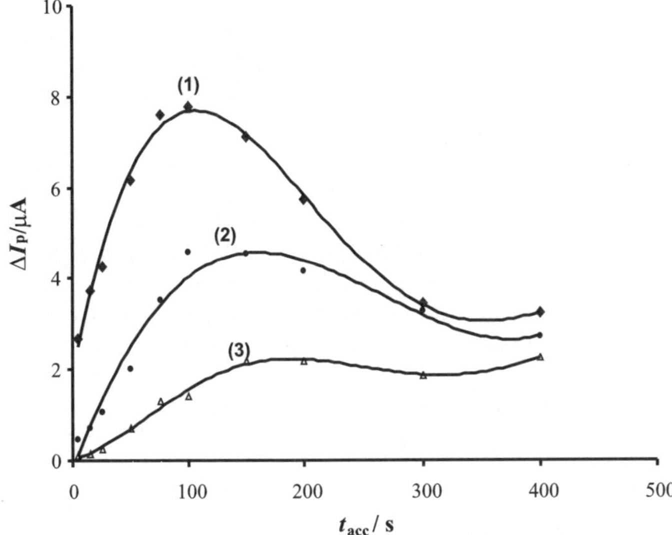

Fig. 2.56 Dependence of the net peak current of probucol on the accumulation time. The amount of acetonitrile in the supporting electrolyte is: 0 (*1*), 3 (*2*) and 6% (*v/v*) (*3*). The other experimental conditions are the same as for Fig. 2.55 (with permission from [85])

Mo(VI) at the mercury surface [95]. The reduction of the Mo(VI) complex proceeds as a one-electron surface electrode process. Due to the complex molecular structure of the deposited compound, attractive interactions emerge within the adsorbed film. The quasireversible maxima have been measured for this system at three distinct accumulation times. The position of the maximum varies with the surface coverage according to the following equation:

$$\ln(f_{\max}/\mathrm{Hz}) = -0.502\,\Theta + 2.820\,(\mathrm{R} = 0.9944) \tag{2.107}$$

In contrast to probucol, the slope of the above line clearly indicates the presence of attractive interactions characterized by a Frumkin interaction parameter of $a = 0.251$ [105].

2.5.3 Surface Electrode Reactions Coupled with Chemical Reactions

In this section the following electrode mechanisms are considered: (a) surface electrode reaction preceded by a reversible chemical reaction [89] ($C_r E$, (2.108) and (2.109)), (b) surface electrode reaction followed by an irreversible chemical reaction [86] (EC_i, (2.109) and (2.110)), and (c) surface catalytic electrode mechanism [87] (EC_i', (2.109 and 2.111)).

$$Y_{(ads)} \underset{k_b}{\overset{k_f}{\rightleftharpoons}} R_{(ads)} \tag{2.108}$$

$$R_{(ads)} \rightleftharpoons O_{(ads)} + ne^- \tag{2.109}$$

$$O_{(ads)} \xrightarrow{k_f} Y_{(ads)} \tag{2.110}$$

$$O_{(ads)} \xrightarrow{k_f} R_{(ads)} \tag{2.111}$$

Similar to the diffusion controlled C_rE mechanism (Sect. 2.4.1) the preceding chemical reaction (2.108) is characterized by the equilibrium constant $K = \frac{k_f}{k_b}$, where k_f and k_b are the first-order rate constants of the forward and backward chemical reactions, respectively. The surface C_rE mechanism is represented by (2.92) and the following differential equations:

$$\frac{d\Gamma_Y}{dt} = k_b\Gamma_R - k_f\Gamma_Y \tag{2.112}$$

$$\frac{d\Gamma_R}{dt} = -\frac{I}{nFA} - k_b\Gamma_R + k_f\Gamma_Y \tag{2.113}$$

The initial conditions are analogous to those for a diffusion controlled C_rE mechanism (Sect. 2.4.1). The only difference is that all species involved in the surface mechanism are immobilized on the electrode surface and characterized by their surface concentrations, instead of volume concentrations used for diffusion controlled C_rE mechanism. In the course of the voltammetric experiment, the following condition holds:

$$t > 0: \quad \Gamma_Y + \Gamma_R + \Gamma_O = \Gamma^* \tag{2.114}$$

The solution for the surface concentration of $R_{(ads)}$ is:

$$\Gamma_R = \frac{K}{1+K}\left(\Gamma^* - \int_0^t \frac{I(\tau)}{nFA}\,d\tau\right) + \frac{1}{k(1+K)}\int_0^t \frac{I(\tau)}{nFA}e^{-k(t-\tau)}\,d\tau \tag{2.115}$$

where $k = k_f + k_b$. The solution for the surface concentration of $O_{(ads)}$ is given by (2.94). Combining (2.115) and (2.94) with the kinetic equation (2.95), the following integral equation is obtained:

$$\frac{I}{nFA} = k_{sur}e^{\alpha_a\varphi}\left[\frac{K}{1+K}\Gamma^* - \frac{K(1+e^{-\varphi})}{1+K}\int_0^t \frac{I(\tau)}{nFA}\,d\tau\right.$$

$$\left. + \frac{1}{k(1+K)}\int_0^t \frac{I(\tau)}{nFA}e^{-k(t-\tau)}\,d\tau\right] \tag{2.116}$$

The numerical solution is:

$$\Psi_m = \frac{\omega\, e^{\alpha_a \varphi_m}\left[\frac{K}{1+K}\left(1-\frac{1}{50}\sum_{j=1}^{m-1}\Psi_j\right) + \frac{1}{\varepsilon(1+K)}\sum_{j=1}^{m-1}\Psi_j M'_{m-j+1} - \frac{e^{-\varphi_m}}{50}\sum_{j=1}^{m-1}\Psi_j\right]}{1-\omega\, e^{\alpha_a \varphi_m}\left(-\frac{K}{50(1+K)} + \frac{M'_1}{\varepsilon(1+K)} - \frac{e^{-\varphi_m}}{50}\right)}$$

$$(2.117)$$

Here, the numerical integration factor is $M'_m = e^{-\frac{\varepsilon}{50}(m)} - e^{-\frac{\varepsilon}{50}(m-1)}$ and $\varepsilon = \frac{k}{f}$ is the chemical dimensionless kinetic parameter, which is the same as for diffusion controlled $C_r E$ reaction (Sect. 2.4.1). The dimensionless current Ψ and the electrode kinetic parameter ω are the same as for the simple surface electrode reaction (Sect. 2.5.1).

For the surface EC_i mechanism, the variation of the Γ_O is described by the following equation:

$$\frac{d\Gamma_O}{dt} = \frac{I}{nFA} - k_f \Gamma_O, \tag{2.118}$$

whereas the variation of Γ_R with time is represented by (2.91). The initial condition for $t = 0$ is

$$t = 0: \Gamma_R = \Gamma^*, \quad \Gamma_O = 0, \quad \Gamma_Y = 0 \tag{2.119}$$

and the boundary conditions for $t > 0$ is given by (2.114). The solution for Γ_O is:

$$\Gamma_O = \int_0^t \frac{I(\tau)}{nFA}\, e^{-k_f(t-\tau)}\, d\tau, \tag{2.120}$$

whereas the solution for Γ_R is given by (2.93). Substituting (2.93) and (2.120) into (2.95) one obtains:

$$\frac{I}{nFA} = k_{sur}\, e^{\alpha_a \varphi}\left[\Gamma^* - \int_0^t \frac{I(\tau)}{nFA}\, d\tau - e^{-\varphi}\int_0^t \frac{I(\tau)}{nFA}\, e^{-k_f(t-\tau)}\, d\tau\right] \tag{2.121}$$

The numerical solution is:

$$\Psi_m = \frac{\omega\, e^{\alpha_a \varphi_m}\left(1 - \frac{1}{50}\sum_{j=1}^{m-1}\Psi_j - \frac{e^{-\varphi_m}}{\varepsilon}\sum_{j=1}^{m-1}\Psi_j M'_{m-j+1}\right)}{1 + \omega\, e^{\alpha_a \varphi_m}\left(\frac{1}{50} + \frac{e^{-\varphi_m} M'_1}{\varepsilon}\right)} \tag{2.122}$$

Here $\varepsilon = \frac{k_f}{f}$ is the chemical kinetic parameter.

For the surface catalytic mechanism EC'_i, the variation of Γ_R is given by:

$$\frac{d\Gamma_R}{dt} = \frac{I}{nFA} + k_f \Gamma_O \tag{2.123}$$

whereas the variation of Γ_O is given by (2.118). The initial and boundary conditions are identical as for the simple surface reaction (Eqs. 2.88 and 2.89). The solution for Γ_R is:

$$\Gamma_R = \Gamma^* - \int_0^t \frac{I(\tau)}{nFA}\, e^{-k_f(t-\tau)}\, d\tau \tag{2.124}$$

and the solution for Γ_O is given by (2.120). Substituting (2.120) and (2.124) into (2.95) one obtains the following integral equation:

$$\frac{I}{nFA} = k_{\mathrm{sur}}\, e^{\alpha_a \varphi} \left[\Gamma^* - \left(1 + e^{-\varphi}\right) \int_0^t \frac{I(\tau)}{nFA}\, e^{-k_f(t-\tau)}\, d\tau \right] \tag{2.125}$$

The numerical solution is:

$$\Psi_m = \frac{\omega\, e^{\alpha_a \varphi_m} \left(1 - \frac{1 + e^{-\varphi_m}}{\varepsilon} \sum_{j=1}^{m-1} \Psi_j M'_{m-j+1}\right)}{1 + \frac{\omega M'_1\, e^{\alpha_a \varphi_m}}{\varepsilon}\left(1 + e^{-\varphi_m}\right)} \tag{2.126}$$

All parameters in the above equations have the same meaning as in case of the surface EC mechanism.

Parameters representing the effect of the chemical reactions, i.e., K and ε, are identically defined as for corresponding mechanisms of a dissolved redox couple (Sect. 2.4); hence their influence on the voltammetric response is rather similar as for the latter mechanisms. For these reasons, in the following part only the unique voltammetric properties of the surface electrode mechanisms coupled with chemical reactions will be addressed.

As for the simple surface reaction (Chap. 2.5.1), all of the aforementioned mechanisms feature the quasireversible maximum. In the theoretical analysis, the quasireversible maximum can be constructed by varying the electrode kinetic parameter ω, while keeping constant the chemical kinetic parameter ε. The position of the theoretical quasireversible maximum is unaffected by the preceding or follow-up chemical reactions. However, in the experimental analysis, the situation can be much more complicated, as the variation of the frequency affects simultaneously both the chemical and electrode kinetic parameters. On the other hand, the splitting of the net peak is highly sensitive to both preceding or following chemical reactions. Shown in Fig. 2.57 is the evolution of the response of the $C_r E$ mechanism for different values of the equilibrium constant. The equilibrium constant affects the splitting only if $\log(\varepsilon) \geq 1$. Under these conditions, the preceding chemical reaction is fast, and the surface concentration of the electroactive reactant Γ_R is predominantly controlled by the thermodynamic parameter K. By controlling the surface concentration Γ_R, the equilibrium constant affects the rate of the electrode reaction, thus influencing the splitting of the net peak. For $\log(K) \geq 2$, the split net peaks are identical with those of a simple surface electrode reaction. Within the interval $-1 < \log(K) < 2$, the potential separation between the split SW peaks decreases by decreasing of K, and the splitting finally vanishes for $\log(K) \leq -1$. When K is large enough ($\log(K) \geq 1$), the surface concentration of Γ_R on the time scale of a single potential pulse is affected by the rate of the preceding chemical reaction. For these reasons, for $\log(K) \geq 1$, the splitting is sensitive to the chemical kinetic parameter ε. The overall effect of ε to the splitting is similar to that of the equilibrium constant K.

Figure 2.58 shows the effect of the chemical kinetic parameter ε on the split SW peaks for surface EC_i mechanism. The peak at more positive potentials, cor-

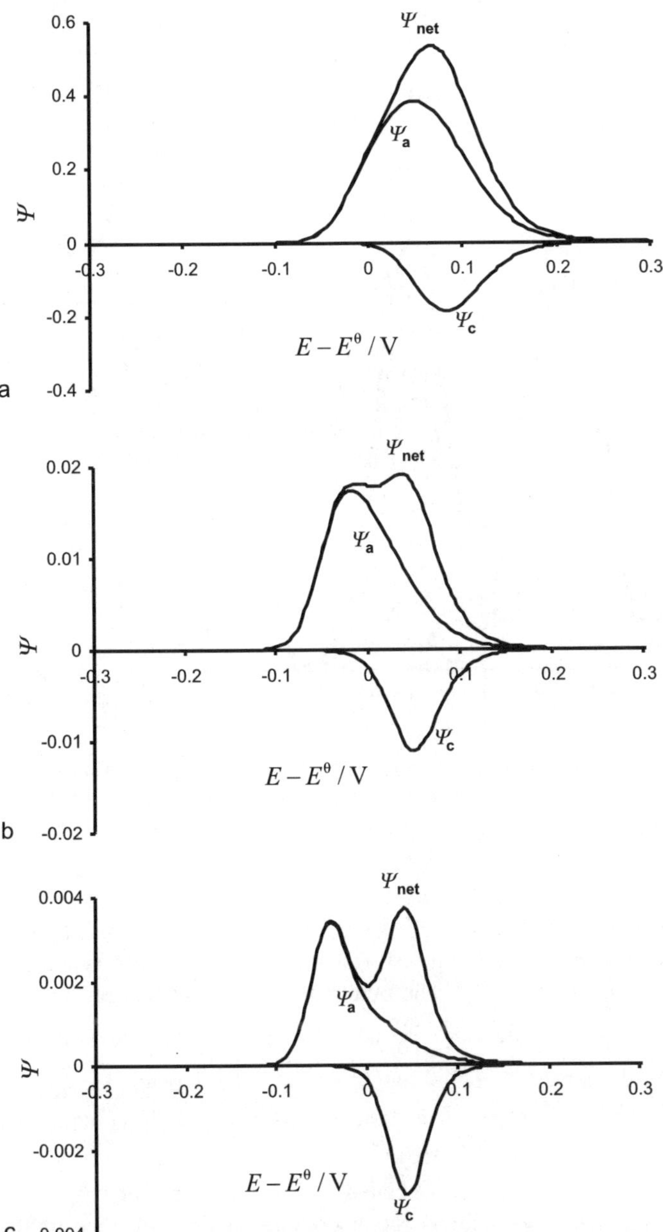

Fig. 2.57a–c C_rE mechanism. Theoretical voltammograms simulated for $K = 0.1$ (**a**); 10 (**b**) and 100 (**c**). The other conditions of the simulations are: $\omega = 10$, $\varepsilon = 10$, $\alpha_a = 0.5$, $nE_{sw} = 50\,\text{mV}$, $\Delta E = 5\,\text{mV}$. *Curves* (Ψ_a), (Ψ_c) and (Ψ_{net}) correspond to the anodic, cathodic and net component of the SW response

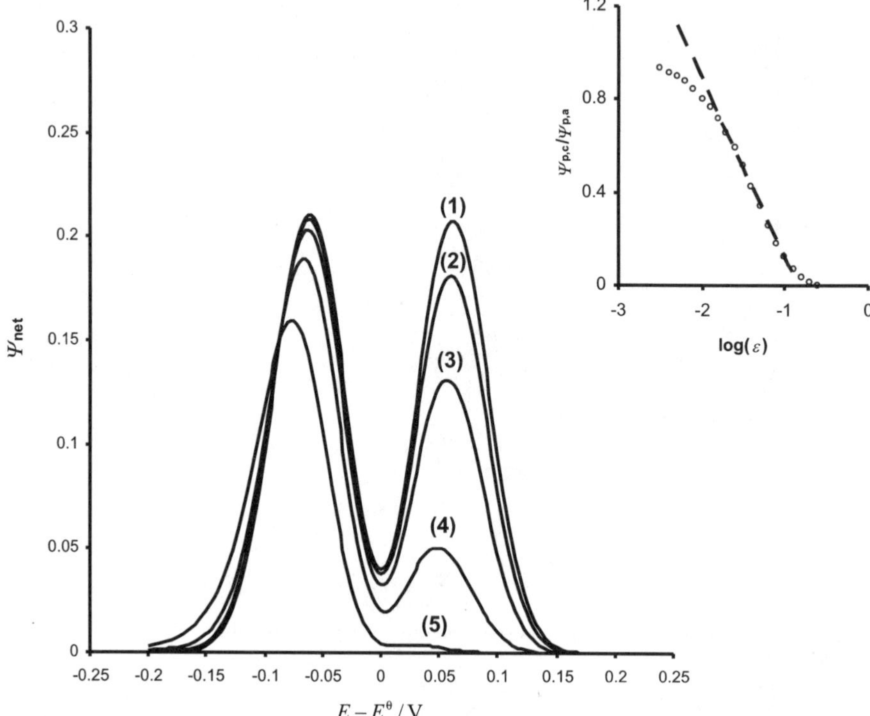

Fig. 2.58 EC$_i$ mechanism. Theoretical net voltammograms simulated for $\log(\varepsilon) = -3$ (*1*); -2 (*2*); -1.5 (*3*); -1 (*4*) and -0.5 (*5*). The other conditions of the simulations are: $\omega = 3$, $\alpha_a = 0.5$, $nE_{sw} = 80$ mV, $\Delta E = 5$ mV. The *inset* shows the dependence of the ratio $\frac{\Psi_{p,c}}{\Psi_{p,a}}$ on $\log(\varepsilon)$ for $\omega = 5$

responding to the reduction process, decreases proportionally to the increasing rate of the follow-up chemical reaction that transforms the electroactive product $O_{(ads)}$ to the final electroinactive form $Y_{(ads)}$. The potential separation between the split peaks is unaltered by ε, whereas the ratio $\frac{\Psi_{p,c}}{\Psi_{p,a}}$ depends sigmoidally on $\log(\varepsilon)$ (inset of Fig. 2.58). Within the interval $-1.5 \leq \log(\varepsilon) \leq -0.5$, the dependence $\frac{\Psi_{p,c}}{\Psi_{p,a}}$ vs. $\log(\varepsilon)$ can be approximated with the linear function associated with the equation:
$$\frac{\Psi_{p,c}}{\Psi_{p,a}} = -0.7682\log(\varepsilon) - 0.6505 \ (R = 0.996).$$

The surface catalytic mechanism is associated with the splitting for $\varepsilon \leq 2$ (see Fig. 2.59). The voltammetric behavior is rather complex under influence of the chemical kinetic parameter. The peak at more negative potential, corresponding to the oxidation process, diminishes slightly with increasing the catalytic parameter. At the first glance, this is unexpected behavior for an oxidative catalytic mechanism. This effect is opposite compared to the EC$_i$ mechanism. The potential separation between the split SW peaks does not depend significantly on the catalytic parameter. Therefore, the splitting of the SW response can be used only as a qualitative indica-

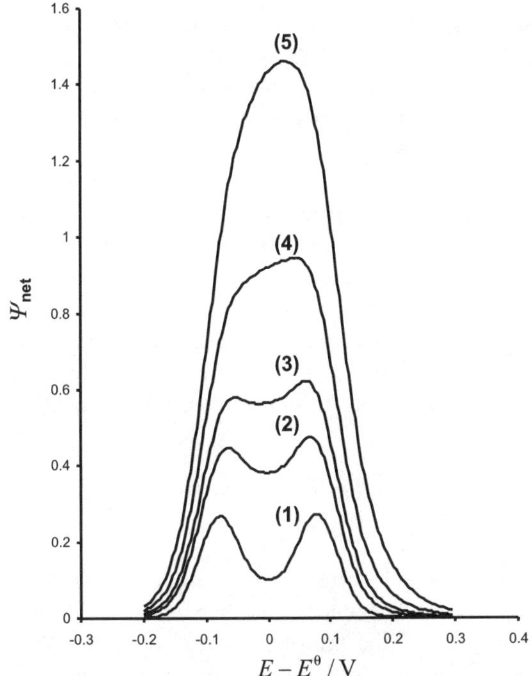

Fig. 2.59 EC$'_\mathrm{i}$ catalytic mechanism. Theoretical net voltammograms simulated for $\varepsilon = 0.1$ (*1*); 0.4 (*2*); 0.6 (*3*); 1 (*4*) and 1.6 (*5*). The other conditions of the simulations are: $\omega = 3$, $\alpha_\mathrm{a} = 0.5$, $nE_\mathrm{sw} = 100\,\mathrm{mV}$, $\Delta E = 5\,\mathrm{mV}$

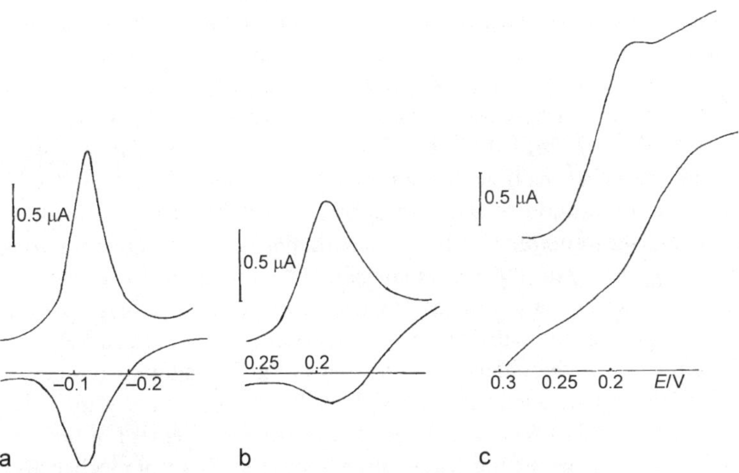

Fig. 2.60 The cathodic and the anodic components of the responses of 1×10^{-4} mol/L solution of azobenzene in 1 mol/L KNO_3 buffered with an acetate buffer to pH $= 4.6$ (**a**), 1 mol/L HNO_3 (**b**), and 2 mol/L HNO_3 (**c**). The other experimental conditions are: $f = 50\,\mathrm{Hz}$, $E_\mathrm{sw} = 25\,\mathrm{mV}$, $\Delta E = 2\,\mathrm{mV}$, $t_\mathrm{acc} = 30\,\mathrm{s}$ (stirred solution), $t_\mathrm{eq} = 5\,\mathrm{s}$, $E_\mathrm{acc} = 0.3\,\mathrm{V}$ (reprinted from [86]; Croat Chem Acta 73:305)

tor to recognize the surface catalytic mechanism; however it can not be exploited as an identifiable feature for estimation of the catalytic rate constant.

Theoretical studies of the surface electrode reactions coupled with chemical reactions have been verified by the voltammetric features of azobenzene [86, 87] and molybdenum complexes [108] adsorbed on the mercury electrode surface. The theory for surface C_rE mechanism is not rigorously verified by a model experimental system. The electrode reaction of azobenzene at the HMDE is a well-known model for a surface process involving two-electron two-proton reduction to hydrazobenzene (2.127). Both components of the redox couple are strongly immobilized on the electrode surface, thus satisfying most of the criteria for a surface process in the absence of significant interactions between the adsorbed species. Hydrazobenzene, being a reduction product of azobenzene, in a strong acidic medium undergoes intramolecular, chemically irreversible rearrangement to electroinactive benzidine ($4',4'$-dyaminobyphenil) (2.128).

$$C_6H_5-N=N-C_6H_{5(ads)} + 2H^+ + 2e^- \rightleftarrows C_6H_5-NH-NH-C_6H_{5(ads)} \qquad (2.127)$$

$$C_6H_5-NH-NH-C_6H_{5(ads)} \rightarrow H_2N-C_6H_4-C_6H_4-NH_{2(ads)} \qquad (2.128)$$

Therefore, in strong acidic medium the simple surface redox reaction of azobenzene turns into a surface EC_i mechanism. The rate of the benzidine rearrangement is proportional to the concentration of the protons in the supporting electrolyte. During the experimental work, the rate of the follow-up chemical reaction can be readily controlled, matching the amount of the acid in the supporting electrolyte. Hence, the anodic branch of the SW response recorded in $1\,mol/L$ HNO_3 is considerably diminished, comparing to the corresponding branch recorded in $1\,mol/L$ KNO_3 (Fig. 2.60). The rate of the benzidine rearrangement in $2\,mol/L$ HNO_3 is sufficiently rapid to transform all the amount of the electrochemically formed hydrazobenzene in benzidine, and therefore the overall electrode reaction appears totally irreversible (Fig. 2.60c). When the SW frequency was increased up to $E_{sw} = 100\,mV$, the SW response of azobenzene recorded in $1\,mol/L$ KNO_3 consisted of two peaks as a result of the splitting of the SW peak [86]. In the presence of nitric acid the peak positioned at more negative potentials (anodic peak) diminishes. If the concentration of the acid is larger than $0.04\,mol/L$, the spitting of the peak vanishes completely. If the experiment is carried out at a constant concentration of the acid, the particular value of the chemical parameter ε should be changed by the variation of the frequency. Figure 2.61 shows how the frequency of the signal changes the apparent reversibility of the electrode process due to its influence to the chemical parameter ε. At the minimal frequency $f = 10\,Hz$, the chemical parameter ε has the highest value; consequently, the electrode reaction appears totally irreversible (Fig. 2.61a). An enhancement of the frequency resulted in an increase of the anodic branch of the SW response because the chemical parameter ε was diminished (Fig. 2.61b). When the frequency is increased up to $f = 200\,Hz$, the influence of the chemical parameter is negligible and the redox reaction of azobenzene appears chemically reversible (Fig. 2.61c). Further increase of the frequency can affect the electrochemical reversibility of the

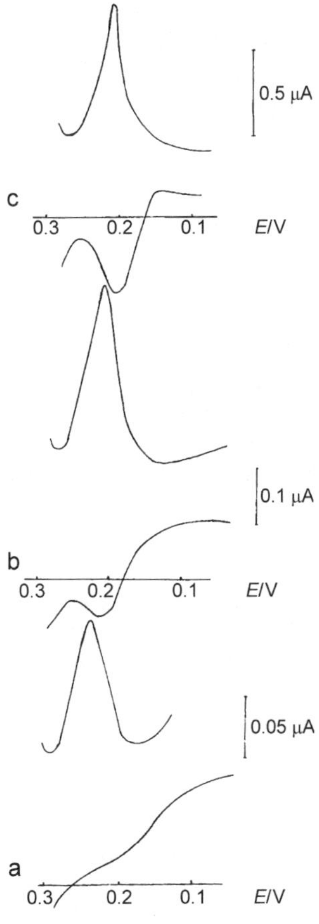

Fig. 2.61 The effect of the frequency on the cathodic and anodic components of the SW response of azobenzene recorded in 1 mol/L HNO_3. The frequency is $f = 10$ (**a**); 50 (**b**) and 200 Hz (**c**). All other conditions are the same as for Fig. 2.60 (reprinted from [86]; Croat Chem Acta 73:305)

redox reaction only through the kinetic parameter $\omega = k_{sur}/f$. If the frequency of the signal is increased above 200 Hz, the redox reaction of azobenzene gradually becomes quasireversible. Due to the effect of the quasireversible maximum, the ratio $\Delta I_p/f$ commences increasing, reaching a maximum value for $f = 600$ Hz (Fig. 2.62).

The theory for surface EC'_i catalytic mechanism was illustrated by the experiments with azobenzene in the presence of hydrogen peroxide as an oxidizing agent [87] and Mo(VI)-mandelic acid complex in the presence of chlorate, bromate and hydrogen peroxide [108]. The oxidizing agents transform the reduction product, i.e., hydrazobenzene or Mo(V)-mandelic acid complex, back to the initial reactant, completing the EC'_i reaction scheme. In an acetate buffer at pH $= 4.2$, the standard rate constant of azobenzene was estimated to be $k_{sur} = 12\ s^{-1}$, whereas the second-order rate constant of hydrazobenzene with hydrogen peroxide is $k_{f,r} = 2.24 \times 10^4\ s^{-1}\,mol^{-1}\,L$. Mo(VI) creates a stable surface active complex with mandelic acid, undergoing a one-electron reduction [108]. In the presence of chlorate or bro-

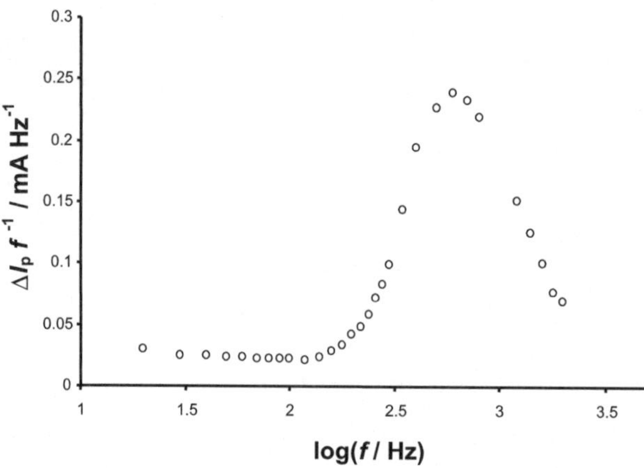

Fig. 2.62 The quasireversible maximum of azobenzene recorded in 1 mol/L HNO_3. All other conditions are the same as for Fig. 2.61 (reprinted from [86]; Croat Chem Acta 73:305)

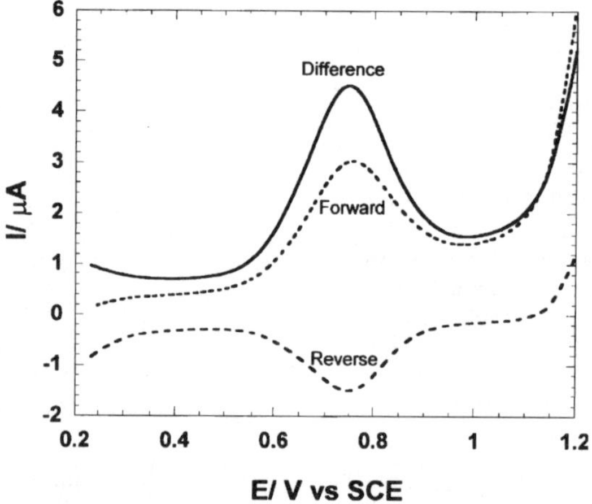

Fig. 2.63 Voltammetric response of $[Ru(bpy)_2\text{-}PVP]^{2+}$ film on pyrolytic graphite electrode recorded at f 5 Hz in 20 mmol/L acetate buffer at pH 5.5 (reprinted from [109] with permission)

mate ions, or hydrogen peroxide, the electrode reaction is transposed into catalytic mechanism. Utilizing the splitting of the net peak and the quasireversible maximum, the standard rate constant of Mo(VI)-mandelic acid system was estimated to be $k_{sur} = 150 \pm 5\,s^{-1}$. By fitting the experimental and theoretical results, the following catalytic rate constants have been estimated: $(8.0 \pm 0.5) \times 10^4\,s^{-1}\,mol^{-1}\,L$, $(1.0 \pm 0.1) \times 10^5\,s^{-1}\,mol^{-1}\,L$, and $(3.2 \pm 0.1) \times 10^6\,s^{-1}\,mol^{-1}\,L$, for hydrogen peroxide, chlorate, and bromate, respectively.

An important example for the surface catalytic mechanism is the catalytic oxidation of DNA at pyrolytic graphite electrode modified with adsorbed poly(4-vinilpyridine) (PVP) with attached $Ru(bpy)_2^{2+}$ (bpy = 2, 2$'$-bipyridyl) [109]. At low

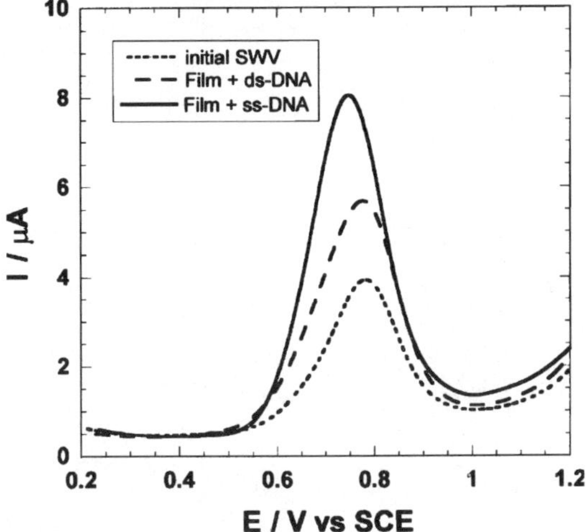

Fig. 2.64 Catalytic SW voltammetric response at 5 Hz of $[Ru(bpy)_2\text{-}PVP]^{2+}$ film at pH 5.5 to equivalent amounts (0.5 mg/mL) of ss- and ds-DNA (reprinted from [109] with permission)

frequencies, these electrodes give reversible voltammetry due to the one electron oxidation of immobilized $Ru(bpy)_2^{2+}$ complex to $Ru(bpy)_2^{3+}$, as shown in Fig. 2.63. These electrodes give catalytic response to poly(guanilic) acid and DNA caused by catalytic oxidation of guanine moieties in these polynucleotides. The polymer film $Ru(bpy)_2^{2+}$–PVP is sensitive to DNA hybridization and damage, as the ss-DNA gave twice the SW catalytic current compared to an equivalent amount of ds-DNA (Fig. 2.64).

2.5.4 Two-Step Surface Electrode Reaction

In the theory of SWV, two different types of surface EE mechanisms have been treated [91, 92]. O'Dea et al. [91] considered a mechanism in which the first redox step was chemically reversible, whereas the second one was a totally irreversible process. In the succeeding study [91], a more general case has been treated consisting of two quasireversible redox transformations, as indicated by (2.129):

$$A_{(ads)} \rightleftarrows B_{(ads)} + n_1\, e^- \rightleftarrows C_{(ads)} + n_2\, e^- \tag{2.129}$$

The first and the second redox reactions are characterized by the distinct standard potentials $E_{A/B}^{\theta}$ and $E_{B/C}^{\theta}$, and different number of exchanged electrons, n_1 and n_2.

The electrode mechanism is represented by the following mathematical model:

$$\frac{d\Gamma_A}{dt} = -\frac{I_1}{n_1 FA} \tag{2.130}$$

$$\frac{d\Gamma_B}{dt} = \frac{I_1}{n_1 FA} - \frac{I_2}{n_2 FA} \tag{2.131}$$

$$\frac{d\Gamma_C}{dt} = \frac{I_2}{n_2 FA} \tag{2.132}$$

$$t = 0: \quad \Gamma_A = \Gamma^*, \quad \Gamma_B = \Gamma_C = 0 \tag{2.133}$$

$$t > 0: \quad \Gamma_A + \Gamma_B + \Gamma_C = \Gamma^* \tag{2.134}$$

$$t > 0: \quad \frac{I_1}{n_1 FA} = k_{\mathrm{sur},1}\, e^{\alpha_{a,1}\varphi_1} \left(\Gamma_A - e^{-\varphi_1}\Gamma_B \right) \tag{2.135}$$

$$\frac{I_2}{n_2 FA} = k_{\mathrm{sur},2}\, e^{\alpha_{a,2}\varphi_2} \left(\Gamma_B - e^{-\varphi_2}\Gamma_C \right) \tag{2.136}$$

As can be seen from the model, each redox reaction is attributed with different set of kinetic parameters. The current contributions are designated with I_1 and I_2. As for the ECE mechanism considered in Sect. 2.4.3, the total current that can be experimentally observed is a sum of distinct current contributions, $I = I_1 + I_2$.

Solutions for the surface concentration of the electroactive species are given by:

$$\Gamma_A = \Gamma^* - \int_0^t \frac{I_1(\tau)}{n_1 FA}\, d\tau \tag{2.137}$$

$$\Gamma_B = \int_0^t \frac{I_1(\tau)}{n_1 FA}\, d\tau - \int_0^t \frac{I_2(\tau)}{n_2 FA}\, d\tau \tag{2.138}$$

$$\Gamma_C = \int_0^t \frac{I_2(\tau)}{n_2 FA}\, d\tau \tag{2.139}$$

Combining (2.137) and (2.138) with kinetic equation (2.135), and (2.138) and (2.139) with (2.136), integral equations are readily obtained as general solutions for each redox step. The numerical solution is represented by the following set of recursive formulas:

$$\Psi_{1,m} = \frac{\omega_1\, e^{\alpha_{a,1}\varphi_{1,m}} \left[1 - \frac{(1+e^{-\varphi_{1,m}})}{50} \sum_{j=1}^{m-1} \Psi_{1,j} + \frac{e^{-\varphi_{1,m}}}{50} \left(\Psi_{2,m} + \sum_{j=1}^{m-1} \Psi_{2,j} \right) \right]}{1 + \frac{\omega_1\, e^{\alpha_{a,1}\varphi_{1,m}}}{50}\left(1 + e^{-\varphi_{1,m}}\right)} \tag{2.140}$$

$$\Psi_{2,m} = \frac{\frac{\omega_2\, e^{\alpha_{a,2}\varphi_{2,m}}}{50} \left[\Psi_{1,m} + \sum_{j=1}^{m-1} \Psi_{1,j} - (1+e^{-\varphi_{2,m}}) \sum_{j=1}^{m-1} \Psi_{2,j} \right]}{1 + \frac{\omega_2\, e^{\alpha_{a,2}\varphi_{2,m}}}{50}\left(1 + e^{-\varphi_{2,m}}\right)} \tag{2.141}$$

where $\Psi_1 = \frac{I_1}{n_1 FA\Gamma^* f}$ and $\Psi_2 = \frac{I_2}{n_2 FA\Gamma^* f}$. To total dimensionless current is $\Psi = \Psi_1 + \Psi_2$, and $\omega_1 = \frac{k_{\mathrm{sur},1}}{f}$ and $\omega_2 = \frac{k_{\mathrm{sur},2}}{f}$ are dimensionless electrode kinetic parameters for each redox reaction.

Two-step surface electrode mechanism generates a wide variety of voltammetric profiles depending mainly on the ratio ω_1/ω_2 and the difference between the standard potentials, $\Delta E^\theta = E^\theta_{B/C} - E^\theta_{A/B}$ (Fig. 2.65). Generally, depending on ΔE^θ, three different cases can be defined: (i) the second redox step occurs at potentials at least 150 mV more positive than the the first step, or $\Delta E^\theta \geq 150\,\mathrm{mV}$; ($ii$): the second redox step occurs at either equal or more negative potential, $\Delta E^\theta \leq 0\,\mathrm{mV}$; and ($iii$) an intermediate case where $0 < (\Delta E^\theta/\mathrm{mV}) < 150$. The first case ($i$) corresponds to the situation when the second redox step is thermodynamically less favourable, occurring at more positive potentials than the first one. Under these conditions, the two redox reactions behave virtually as independent processes, and the overall response consists of two well-separated peaks (curves 4 and 5 in Fig. 2.65). The voltammetric features of each process can be satisfactorily explained on the basis of the theory presented in Sect. 2.5.1 referring to the simple surface electrode reaction. In case (ii), the thermodynamic conditions for electrochemical conversion of the intermediate B to the final product C are fulfilled immediately after formation of B via the first redox reaction. This is an exceptionally important case, since it can be regarded as a possible pathway of many two-electron quasireversible surface redox reactions. The response consists of a single SW peak, the features of which are determined by kinetic parameters ω_1 and ω_2 and electron-transfer coefficients $\alpha_{a,1}$ and $\alpha_{a,2}$. Figure 2.66 illustrates the effect of ω_2 for several constant values of

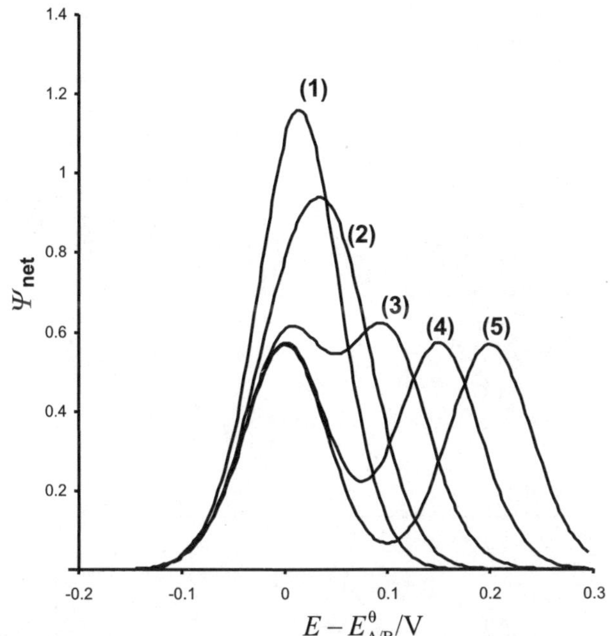

Ψ_{net}

$E - E^\theta_{A/B}/\mathrm{V}$

Fig. 2.65 Theoretical net voltammograms simulated for $\Delta E^\theta = 0$ (*1*); 50 (*2*); 100 (*3*); 150 (*4*) and 200 mV (*5*). The other conditions of the simulations are: $\omega_1 = \omega_2 = 1$, $\alpha_{a,1} = \alpha_{a,2} = 0.5$, $n_1 = n_2 = 1, E_{\text{sw}} = 30\,\mathrm{mV}$, and $\Delta E = 5\,\mathrm{mV}$

ω_1. This analysis corresponds to a comparison of different two-step processes characterized by identical kinetics of the first redox step and different kinetics of the second one. For $\log(\omega_1) = -3.5$ (curve 1 in Fig. 2.66), ω_2 exhibits no influence on the net peak current, whereas for $\log(\omega_1) = -1$ (curve 2 in Fig. 2.66) the effect is very week. Over the interval $\log(\omega_1) \leq -3.5$ limiting conditions are reached and kinetics of the overall reaction is solely controlled by the first redox step, which is slow and electrochemically irreversible.

The second limiting situation appears when the kinetic parameter of the first redox step is very large $(\log(\omega_1) > 0.8)$; hence the kinetics of the overall process is solely determined by the second redox step (curves 5 and 6 in Fig. 2.66). In the intermediate case, $-3.5 \leq \log(\omega_1) \leq 0.8$, the kinetics of the two-step surface reaction is a complex function of the kinetics of both redox steps (curves 3 and 4 in Fig. 2.66). Care must be taken that the foregoing theoretical consideration is not valid for the study of a single two-step electrode reaction, as the kinetic parameters ω_1 and ω_2 cannot be independently varied. In the real experiment, one can affect the electrochemical reversibility of each step by adjusting the signal frequency, which alters simultaneously both kinetic parameters. Changing the frequency from a certain minimal value $f_{\min}$ up to a certain maximal value $f_{\max}$, for $\log(k_{\mathrm{sur},1}/f_{\min}) \leq -3.5$ the limiting situation in which the overall process is controlled by the kinetics of the first redox step is achieved. Whereas, the overall process is controlled solely by the

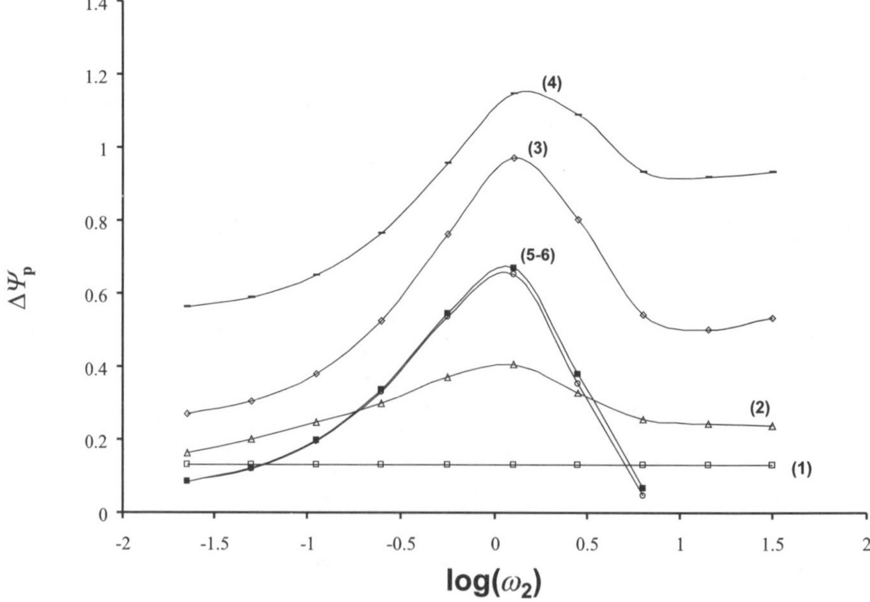

Fig. 2.66 Dependence of the dimensionless peak current $\Delta\Psi_\mathrm{p}$ on the kinetic parameter ω_2 for $\Delta E^\theta = 0\,\mathrm{mV}$. The other kinetic parameter is $\log(\omega_1) = -3.5$ (*1*), -1 (*2*), 0 (*3*), 0.48 (*4*), 0.90 (*5*), and 1 (*6*). The other conditions of the simulations are: $\alpha_{\mathrm{a},1} = \alpha_{\mathrm{a},2} = 0.5$, $n_1 = n_2 = 1$, $\Delta E = 10\,\mathrm{mV}$ and $E_{\mathrm{sw}} = 30\,\mathrm{mV}$ (reprinted from [91]; Croat Chem Acta 76:37)

kinetics of the second redox step for $\log(k_{\mathrm{sur},1}/f_{\max}) \geq 0.8$. A more detailed discussion of the complex intermediate situation, when both redox steps affect the kinetics of the overall process can be found in [91].

In the third case (*iii*), the response of the two-step reaction consists of either a single or two overlapped SW peaks, depending on the ratio of the kinetic parameters ω_1/ω_2 as well as the exact value of ΔE^θ. Figure 2.67 depicts several net responses simulated for $\Delta E^\theta = 100\,\mathrm{mV}$ and different values of the ratio ω_1/ω_2. For $\omega_1/\omega_2 \leq 0.2$ a single peak is observed (Fig. 2.67, curve 1). Increasing the ratio ω_1/ω_2 causes the response to split into two peaks. For $\omega_1/\omega_2 = 0.79$ the potential separation between the overlapped SW peaks is $70\,\mathrm{mV}$ (Fig. 2.67, curve 2). In general, the potential separation increases in proportion to ω_1/ω_2. It is important to note that for $\omega_1/\omega_2 = 50$, the potential separation is $180\,\mathrm{mV}$, which is larger than the actual difference in standard potentials ΔE^θ (see curve 3 in Fig. 2.67). It is worth noting that if the potential separation between the SW peaks is greater than $75\,\mathrm{mV}$, the kinetics of both redox steps may be independently inspected utilizing the quasireversible maximum, or some other method for kinetic measurements.

Two experimental systems have been used to illustrate the theory for two-step surface electrode mechanism. O'Dea et al. [90] studied the reduction of Dimethyl Yellow (4-(dimethylamino)azobenzene) adsorbed on a mercury electrode using the theory for two-step surface process in which the second redox step is totally irreversible. The thermodynamic and kinetic parameters have been derived from a pool of 11 experimental voltammograms with the aid of COOL algorithm for nonlinear least-squares analysis. In Britton–Robinson buffer at pH 6.0 and for a surface concentration of $1.73 \times 10^{-11}\,\mathrm{mol\,cm^{-2}}$, the parameters of the two-step reduction of Dimethyl Yellow are: $E_1^{\theta'} = -0.397 \pm 0.001\,\mathrm{V}$ vs. SCE, $\alpha_{c,1} = 0.43 \pm 0.02$, $k_{\mathrm{sur},1} = 103 \pm 8\,\mathrm{s^{-1}}$, $\alpha_{c,2} = 0.11 \pm 0.04$ and $k_{\mathrm{sur},2}$ (referenced to $E_1^{\theta'}$) $= 11.1 \pm 1.7\,\mathrm{s^{-1}}$. The reduction mechanism of Dimethyl Yellow proceeds according to the scheme:

$$\mathrm{C_6H_5-N=N-C_6H_4-N(CH_3)_2} \underset{E_1^\theta, k_{\mathrm{sur},1}, \alpha_{c,1}}{\overset{2\mathrm{H}^+ + 2\mathrm{e}^-}{\rightleftharpoons}} \mathrm{C_6H_5-NH-NH-C_6H_4-N(CH_3)_2}$$

$$\overset{2\mathrm{H}^+ + 2\mathrm{e}^-}{\underset{E_1^\theta\, k_{\mathrm{sur},2}, \alpha_{c,2}}{\rightarrow}} \mathrm{C_6H_5-NH_2 + NH_2-C_6H_4-N(CH_3)_2} \tag{2.142}$$

The reduction potential of the second redox step overlaps with the potential of the first one, resulting in an overall four-electron four-proton irreversible reduction. The features of the voltammetric response are controlled by the competition between reaction pathways of the hydrazo-form, which can be either reoxidized back to the azo-form or irreversibly reduced to the electroinactive amines.

The second experimental system explored the reduction mechanism of another azo-dye, known as Sudan III (1-(4-phenylazophenylazo)-2-naphthol) [91]. Sudan III contains two azo groups rendering two successive two-electron, two-proton reduction steps at the mercury surface. Figure 2.68 shows a typical SW voltammetric response of Sudan III recorded in a borate buffer at pH 10.00. The first reduction step is chemically reversible, while the second one is irreversible. More importantly, the second reduction step proceeds at potential about $230\,\mathrm{mV}$ more negative than the first one, thus causing a well-separated voltammetric peak. The overall mechanism

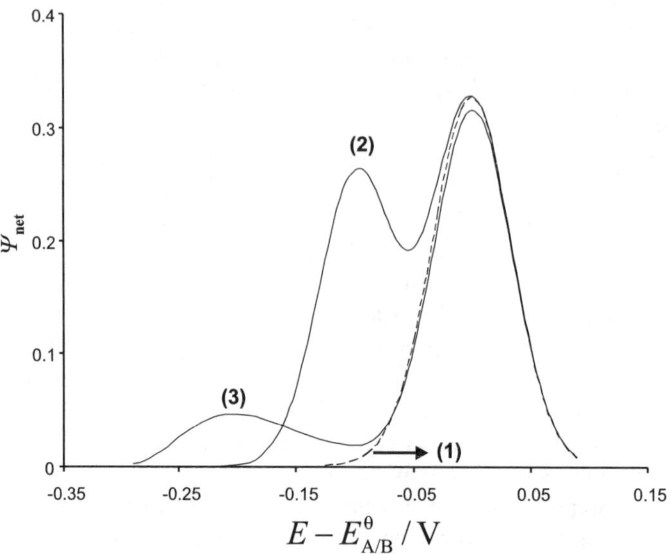

Fig. 2.67 Theoretical net voltammograms of EE reaction simulated for $\Delta E^{\theta} = 100$ mV. The ratio of the kinetic parameters is $\omega_1/\omega_2 = 0.2$ (*1*), 0.7 (*2*) and 50 (*3*). The other conditions of the simulations are: $\alpha_{a,1} = \alpha_{a,2} = 0.5$, $n_1 = n_2 = 1$, $\Delta E = 10$ mV and $E_{sw} = 30$ mV (reprinted from [91]; Croat Chem Acta 76:37)

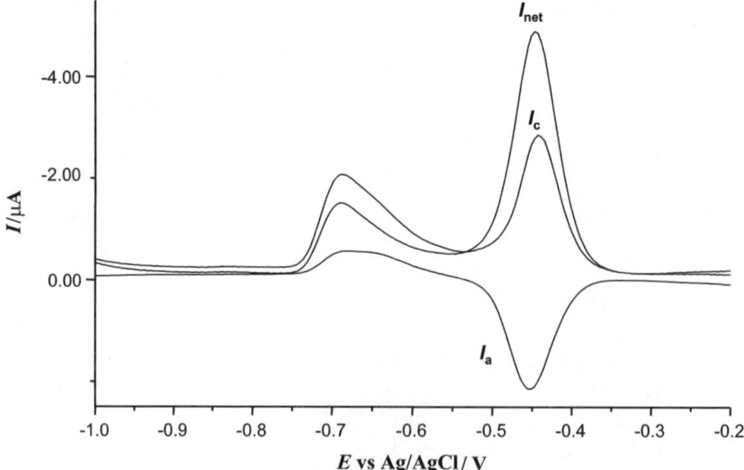

Fig. 2.68 Square-wave voltammogram of 5×10^{-5} mol/L SUDAN III solution recorded in a borate buffer at pH $= 10$. The experimental conditions are: $E_{acc} = -0.2$ V, $t_{acc} = 30$ s, $E_{sw} = 30$ mV, $f = 100$ Hz and $\Delta E = 4$ mV. *Symbols* I_c, I_a, and I_{net} correspond to the cathodic, anodic and net current components of the SW response (reprinted from [91]; Croat Chem Acta 76:37)

corresponds to the case (*i*), when the two redox steps act as independent electrode processes.

2.6 Mixed-Electrode Reactions

2.6.1 Electrode Reactions Coupled with Adsorption of the Reactant and Product of the Electrode Reaction

Frequently one observes mixed-electrode reactions involving adsorption equilibria of electroactive species. Unlike surface electrode processes (Sect. 2.5), where electroactive species are firmly immobilized on the electrode surface and the mass transport is negligible on the time scale of the voltammetric experiment, such mixed reactions produce current signals caused by electrode reactions of both the adsorbed and the dissolved forms of the electroactive species. The theory of SWV of adsorption coupled electrode processes has been extensively developed because of their particular importance in electroanalytical chemistry [79, 92, 110–119]. Electrode mechanisms treated in this chapter are relevant to electrode processes involving (a) anion-induced adsorption of metal ions, (b) electrode reactions of metal complexes with organic ligands, (c) electrode reactions of miscellaneous organic compounds, and (d) various other processes studied in adsorptive stripping voltammetry.

In the following chapter, two general electrode mechanisms are considered. In the first one, only the reactant adsorbs at the electrode surface [92, 110, 111, 114, 115]:

$$R_{(aq)} \rightleftarrows R_{(ads)} \tag{2.143}$$

$$R_{(ads)} \rightleftarrows O_{(aq)} + n\,e^- \tag{2.144}$$

The second mechanism is more general, since both reactant and product adsorbs on the electrode surface [79, 92, 110, 112]:

$$R_{(aq)} \rightleftarrows R_{(ads)} \tag{2.145}$$

$$R_{(ads)} \rightleftarrows O_{(ads)} + n\,e^- \tag{2.146}$$

$$O_{(ads)} \rightleftarrows O_{(aq)} \tag{2.147}$$

The former reaction mechanism involving adsorption of the reactant only can be regarded as a limiting case of latter mechanism. As in the Sect. 2.1 the modeling of adsorption complicated reactions is restricted to the semi-infinite diffusion model at a planar electrode. It is important to emphasize that the adsorption equilibrium is described by a linear adsorption isotherm due to its relevance for electroanalytical methods. The linear, or Henry isotherm, describes the conditions on the electrode when its surface is only partly covered with a submonolayer of electroactive species exhibiting no lateral interactions between them. The surface concentration Γ is much lower than the surface concentration under saturation coverage, Γ_{max}. The surface concentration of the adsorbed electroactive form, i.e., the concentration in the inner Helmholtz plane Γ, is proportional to the concentration of the dissolved form encountered at the outer Helmholtz plane ($c_{x=0}$), i.e., $\Gamma = \beta(c_{x=0})$. The proportionality constant β is the so-called adsorption constant with units of cm. A model dealing with the more complex Frumkin adsorption isotherm at a spherical electrode is also available [120]. For modeling of reaction (2.144), differential equations (1.2) and (1.3) have to be solved under the following initial and boundary conditions:

$$t = 0, \quad x \geq 0: \quad c_R = c_R^*; \quad \Gamma_R = 0; \quad c_O = 0; \tag{2.148}$$

$$t > 0, \quad x \to \infty: \quad c_R \to c_R^*; \quad c_O \to 0 \tag{2.149}$$

$$t > 0, \quad x = 0: \quad \beta_R (c_R)_{x=0} = \Gamma_R \tag{2.150}$$

$$\frac{I}{nFA} = D \left(\frac{\partial c_R}{\partial x} \right)_{x=0} - \frac{d\Gamma_R}{dt} \tag{2.151}$$

$$D \left(\frac{\partial c_O}{\partial x} \right)_{x=0} = -\frac{I}{nFA} \tag{2.152}$$

As follows from the initial condition (2.148) no adsorbed form is present on the electrode surface prior to the voltammetric experiment. The boundary condition (2.151) shows that the current is produced by the flux of the dissolved reactant as well as by its adsorbed form. The solution for the R form at the electrode surface is:

$$(c_R)_{x=0} = c_R^* \left[1 - \exp(a_R^2 t) \operatorname{erfc}(a_R \sqrt{t}) \right]$$

$$- a_R \int_0^t \frac{I(\tau)}{nFA\sqrt{D}} \exp\left[a_R^2(t - \tau) \right] \operatorname{erfc}(a_R \sqrt{t - \tau}) \, d\tau \tag{2.153}$$

where $a_R = \frac{\sqrt{D}}{\beta_R}$ is the auxiliary adsorption parameter. The solution for the product O is given by (2.38). If reaction (2.144) is reversible, (2.153) and (2.38) are combined with the Nernst equation (1.8) to yield an integral equation, whose numerical solution is:

$$\Psi_m = \frac{1 - \exp\left(\frac{\rho_R^2 m}{50} \right) \operatorname{erfc}(\rho_R \sqrt{\frac{m}{50}}) - \sum_{j=1}^{m-1} \Psi_j \left(\frac{2S_{m-j+1}}{\sqrt{50\pi}} + \frac{(R_R)_{m-j+1}}{\rho_R} \right) - \frac{2\exp(-\varphi_m)}{\sqrt{50\pi}} \sum_{j=1}^{m-1} \Psi_j S_{m-j+1}}{\frac{2\exp(-\varphi_m)}{\sqrt{50\pi}} + \frac{2}{\sqrt{50\pi}} + \frac{(R_R)_1}{\rho_R}} \tag{2.154}$$

Here $\rho_R = \frac{1}{\beta_R} \sqrt{\frac{D}{f}}$ is a dimensionless adsorption parameter, $(R_R)_m$ is an integration parameter defined as

$$(R_R)_m = \exp\left(\frac{m\rho_R^2}{50} \right) \operatorname{erfc}\left(\rho_R \sqrt{\frac{m}{50}} \right) - \exp\left(\frac{(m-1)\rho_R^2}{50} \right) \operatorname{erfc}\left(\rho_R \sqrt{\frac{m-1}{50}} \right) \tag{2.155}$$

and S_m is defined as in (2.40).

In the experimental analysis the SW potential scan is preceded by a certain delay period (t_{delay}) to allow the reactant to adsorb on the electrode surface. Besides, the reactant adsorbs additionally in the course of the voltammetric scan starting from the initial (E_s) to the peak potential (E_p). Thus, the total accumulation period (t_{acc})

is defined as:

$$t_{acc} = t_{delay} + \frac{E_p - E_s}{f \Delta E}$$
(2.156)

Note that the product $f \Delta E$ yields the scan rate of the square-wave potential modulation. If the delay period is sufficiently long, the additional adsorption during the potential scan is negligible. Otherwise, the additional adsorption complicates the theoretically expected dependencies, in particular the relationships between the net peak currents and potentials on the frequency [114].

The most critical parameter of reaction (2.144) is the adsorption parameter $\rho_R = \frac{1}{\beta_R} \sqrt{\frac{D}{f}}$. It couples the role of adsorption strength together with the diffusion mass transport on the time scale of potential pulses. If the adsorption is very weak, $\rho_R \geq 1.23$, the response of reaction (2.144) is equivalent to the simple reaction of a dissolved redox couple (2.157).

$$R_{(aq)} \rightleftarrows O_{(aq)} + n e^-$$
(2.157)

Figure 2.69 compares the theoretical responses of an adsorption coupled reaction with the simple reaction of a dissolved redox couple, for a reversible case. Obviously, the adsorption enhances considerably the response, making the oxidation process more difficult. The forward component of reaction (2.144) is a sharp peak, with a lower peak width compared to reaction (2.157). The relative position of the peak potentials of the forward and backward components of the adsorption complicated reaction is inverse compared to simple reaction of a dissolved redox couple. Finally, the peak current of the stripping (forward) component of adsorption coupled reaction is lower than the backward one, the ratio $\frac{I_{p,f}}{I_{p,b}}$ being 0.816. The corresponding value for reaction of a dissolved couple is 1.84. This anomaly is a consequence of the current sampling procedure and immobilization of the reactant, as explained in the Sect. 2.5.1.

Further significant differences between reactions (2.144) and (2.157) can be revealed by varying the signal frequency. The real net peak current of an adsorption complicated reaction is a linear function of f (Fig. 2.70). Recall that for a simple reaction of a dissolved redox couple the net peak current depends linearly on $\sqrt{f}$ (see Sect. 2.1.1). Figure 2.70 shows that for low frequencies and a low delay time, deviations from linearity occur, which are a consequence of the additional adsorption during the potential scan. Hence, rigorously speaking, the real net peak current of adsorption coupled reaction (2.144) is a nonlinear function of f, characterized by the point $\Delta I_p = 0$ for $f = 0$ and an asymptote $\Delta I_p = kf + z$, where the intercept z depends on the delay time and apparently vanishes for $t_{delay} \geq 30$ s. In contrast to reaction (2.157), the peak potential of an adsorption coupled reaction depends linearly on $\log(f)$ with a slope $\frac{\Delta E_p}{\Delta \log(f)} = 2.3 \frac{RT}{2nF}$ for an oxidative mechanism. Note that for a reductive mechanism the slope is identical in absolute value but opposite in the sign.

If reaction (2.144) is quasireversible, two reaction pathways are possible:

$$R_{(aq)} \overset{k_s}{\rightleftarrows} O_{(aq)} + n e^-$$
(2.158)

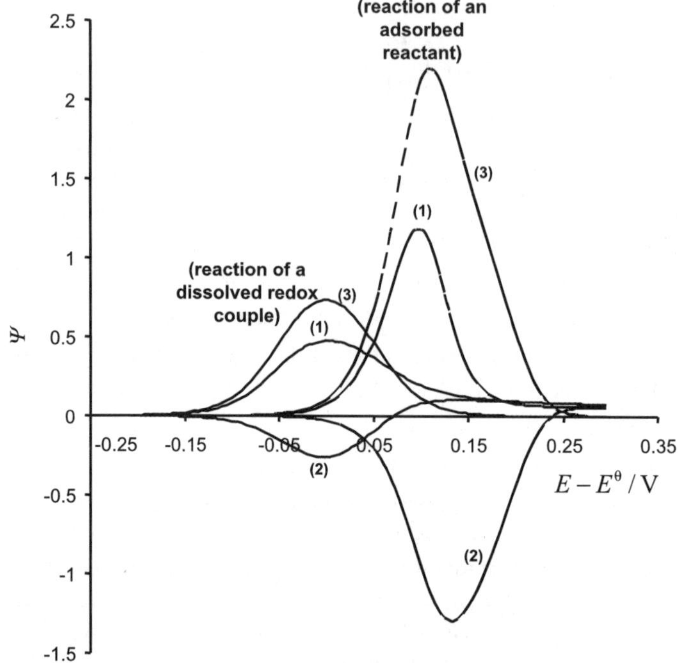

Fig. 2.69 Theoretical voltammograms for a reversible reaction coupled with adsorption of the reactant (2.144) and simple reversible reaction of a dissolved redox couple (2.157). Conditions of the simulations are: $\beta_R = 0.1$ cm, $f = 100$ Hz, $\Delta E = 5$ mV, $nE_{sw} = 50$ mV, $D = 5 \times 10^{-6}$ cm^2 s^{-1}. The numbers *1*, *2*, and *3* designate the forward, backward and net component of the response, respectively

and

$$R_{(ads)} \overset{k_{sur}}{\rightleftarrows} O_{(aq)} + n\,e^- \tag{2.159}$$

The quasireversible reaction (2.158) of a dissolved redox couple is described with the following kinetic equation:

$$\frac{I}{nFA} = k_s \exp(\alpha_a \varphi)\left[(c_R)_{x=0} - \exp(-\varphi)(c_O)_{x=0}\right] \tag{2.160}$$

characterized by a common standard rate constant k_s in units of cm s^{-1}. The quasire-

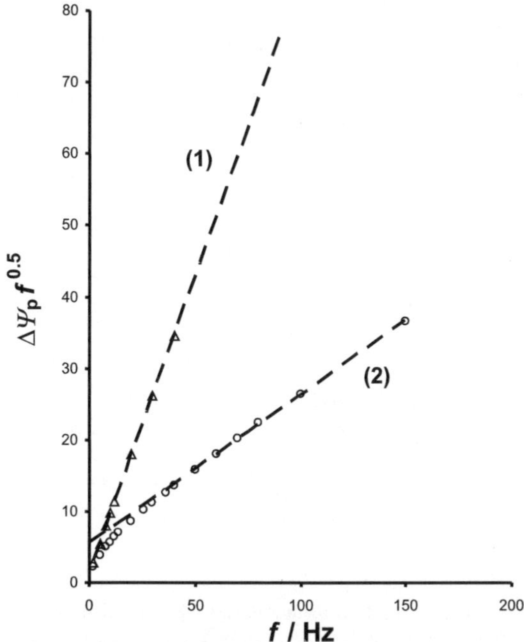

Fig. 2.70 Reversible reaction coupled with adsorption of the reactant. Effect of f on the product $\Delta\Psi_p f^{0.5}$. Conditions of the simulations are: $\beta_R = 0.1$ cm, $E_{acc} = -0.4$ vs. E^{θ}, $\Delta E = 5$ mV, $nE_{sw} = 50$ mV, $D = 5 \times 10^{-6}$ cm^2 s^{-1}, $t_{delay} = 10$ (*1*) and 0.5 s (*2*)

versible surface reaction (2.159) has the surface standard rate constant k_{sur} in units of s^{-1} and the following form of the kinetic equation holds:

$$\frac{I}{nFA} = k_{sur}\exp(\alpha_a\varphi)\left[\Gamma_R - \exp(-\varphi)r_s(c_O)_{x=0}\right] \tag{2.161}$$

where $r_s = 1$ cm is an auxiliary constant. Frequently it is assumed that the surface reaction is faster than the volume one, thus the reaction (2.159) predominates over reaction (2.158). Combining (2.150), (2.153), (2.38) and (2.161), an integral equation is obtained with the following numerical solution:

$$\Psi_m = \frac{\omega\exp(\alpha_a\varphi_m)\left\{\frac{1}{\rho_R}\left[1 - \exp\left(\frac{\rho_R^2 m}{50}\right)\mathrm{erfc}(\rho_R\sqrt{\frac{m}{50}}) - \sum_{j=1}^{m-1}\Psi_j\left(\frac{2S_{m-j+1}}{\sqrt{50\pi}} + \frac{(R_R)_{m-j+1}}{\rho_R}\right)\right] - \frac{2\gamma\exp(-\varphi_m)}{\sqrt{50\pi}}\sum_{j=1}^{m-1}\Psi_j S_{m-j+1}\right\}}{1 + \omega\exp(\alpha_a\varphi_m)\left[\frac{1}{\rho_R}\left(\frac{2}{\sqrt{50\pi}} + \frac{(R_R)_1}{\rho_R}\right) + \frac{2\gamma\exp(-\varphi_m)}{\sqrt{50\pi}}\right]} \tag{2.162}$$

Here, $\omega = \frac{k_{sur}}{f}$ is the electrode kinetic parameter typical for surface electrode processes (see Sect. 2.5.1) and $\gamma = r_s\sqrt{\frac{f}{D}}$ is dimensionless diffusion parameter. The latter parameter represents the influence of the mass transfer of electroactive species.

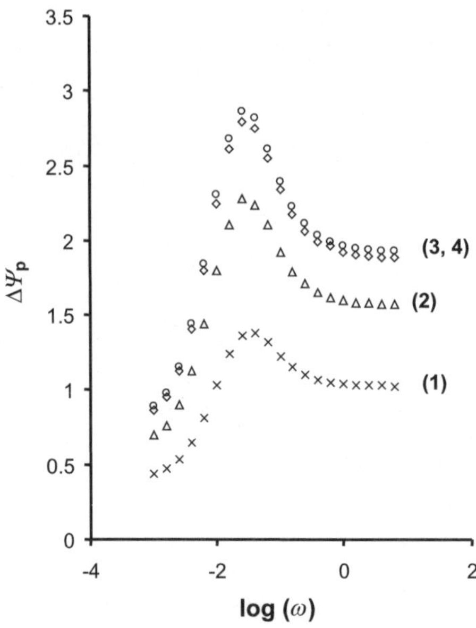

Fig. 2.71 Quasireversible reaction coupled with adsorption of the reactant. Effect of the kinetic parameter ω on $\Delta\Psi_p$. Conditions of the simulations are: $E_{acc} = -0.4$ vs. E^θ, $t_{delay} = 1$ s, $\alpha_a = 0.5$, $\Delta E = 10$ mV, $nE_{sw} = 50$ mV, $D = 5 \times 10^{-6}$ cm^2 s^{-1}, $\beta_R = 0.002$ (*1*); 0.01 (*2*); 0.1 (*3*) and 1 cm (*4*)

The physical meaning of the kinetic parameter ω is identical as for surface electrode reaction (Chap. 2.5.1). The electrochemical reversibility is primarily controlled by ω (Fig. 2.71). The reaction is totally irreversible for $\log(\omega) < -3$ and electrochemically reversible for $\log(\omega) > 1$. Between these intervals, the reaction appears quasireversible, attributed with a quasireversible maximum. Though the absolute net peak current value depends on the adsorption parameter, Fig. 2.71 reveals that the quasireversible interval, together with the position of the maximum, is independent of the adsorption strength. Similar to the surface reactions, the position of the maximum varies with the electron transfer coefficient and the amplitude of the potential modulation [92].

It should be emphasized that the physical meaning of the analysis in Fig. 2.71 corresponds to the comparison of the net peak currents of various electrode reactions characterized by different electrode kinetics. For a single electrode reaction, the electrochemical reversibility will vary by changing the frequency. The net peak current is expected to be a complex function of the frequency, since the latter affects simultaneously all three parameters ω, ρ and γ. Simulations have shown that the $\Delta\Psi_p$ is a non-linear function of f without reaching a maximum. However, the dependence of the ratio $\Delta\Psi_p f^{-0.5}$ vs. $\log(f)$ exhibits a maximum, the position of which is independent of the adsorption strength within the interval $0.01 \leq \beta_R \leq 1$ (Fig. 2.72). Note that the ratio $\Delta\Psi_p f^{-0.5}$ corresponds to the ratio of the real net peak current and the frequency, $\Delta I_p f^{-1}$. This means that the quasireversible maximum

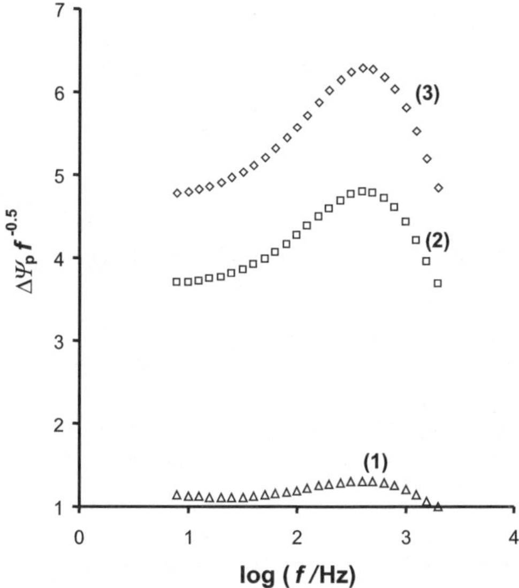

Fig. 2.72 Quasireversible reaction coupled with adsorption of the reactant. Effect of f on the ratio $\Delta\Psi_p f^{-0.5}$. Conditions of the simulations are: $E_{acc} \ll E^\theta$, $k_{sur} = 5\,\mathrm{s}^{-1}$, $\alpha_a = 0.5$, $t_{delay} = 30\,\mathrm{s}$, $\Delta E = 5\,\mathrm{mV}$, $nE_{sw} = 50\,\mathrm{mV}$, $D = 5 \times 10^{-6}\,\mathrm{cm^2\,s^{-1}}$, $\beta_R = 0.01$ (*1*); 0.1 (*2*); and 1 cm (*3*)

can be experimentally found by plotting the ratio $\Delta I_p f^{-1}$ vs. $\log(f)$. The position of the maximum is associated with the critical kinetic parameter $\omega_{max} = 0.0125$ for the conditions corresponding to Fig. 2.72. Hence, fitting the position of the theoretically calculated maximum with the experimentally measured one, the standard rate constant can be estimated.

If the adsorption coupled reaction (2.144) is totally irreversible, the voltammetric complexity is significantly reduced [111, 115]. For the totally irreversible case, the real net peak current is a linear function of the frequency, whereas the peak potentials depends linearly on $\log(f)$ with a slope of $\frac{\Delta E_p}{\Delta \log(f)} = 2.3 \frac{RT}{\alpha_a nF}$. This slope enables estimation of the electron transfer coefficient, provided the number of exchanged electrons is known. Similarly, the same parameter can be inferred from the half-peak width, which is defined as $\Delta E_{p/2} = (63.5 \pm 0.5)/\alpha_a n$ mV.

For modeling of the electrode reaction coupled with adsorption of both forms of the redox couple (2.146), the diffusion equations (1.2) and (1.3) have to be solved for conditions given by (2.148) to (2.152) completed with the following boundary conditions:

$$t = 0, \quad x \geq 0: \quad \Gamma_O = 0 \tag{2.163}$$

$$t > 0, \quad x = 0: \quad \beta_O\,(c_O)_{x=0} = \Gamma_O \tag{2.164}$$

$$D\left(\frac{\partial c_O}{\partial x}\right)_{x=0} = -\frac{I}{nFA} + \frac{\mathrm{d}\Gamma_O}{\mathrm{d}t} \tag{2.165}$$

The solution for the reactant R is given by (2.153), whereas the solution for the product is:

$$(c_O)_{x=0} = a_O \int_0^t \frac{I(\tau)}{nFA\sqrt{D}} \exp\left[a_O^2(t-\tau)\right] \mathrm{erfc}(a_O\sqrt{t-\tau}) \, \mathrm{d}\tau , \qquad (2.166)$$

where $a_O = \frac{\sqrt{D}}{\beta_O}$ is the auxiliary adsorption parameter.

For the reversible case, an integral equation is obtained by combination of (2.153), (2.166), and (1.8). The numerical solution of such integral equation is:

$$\Psi_m = \frac{\begin{aligned}&1 - \exp\left(\frac{\rho_R^2 m}{50}\right) \mathrm{erfc}(\rho_R\sqrt{\tfrac{m}{50}}) - \sum_{j=1}^{m-1} \Psi_j \left(\frac{2S_{m-j+1}}{\sqrt{50\pi}} + \frac{(R_R)_{m-j+1}}{\rho_R}\right)\\[2mm] &\quad - \exp(-\varphi_m) \sum_{j=1}^{m-1} \Psi_j \left(\frac{2S_{m-j+1}}{\sqrt{50\pi}} + \frac{(R_O)_{m-j+1}}{\rho_O}\right)\end{aligned}}{\exp(-\varphi_m)\left(\frac{2}{\sqrt{50\pi}} + \frac{(R_O)_1}{\rho_O}\right) + \frac{2}{\sqrt{50\pi}} + \frac{(R_R)_1}{\rho_R}} \qquad (2.167)$$

Here, $\rho_O = \frac{1}{\beta_O}\sqrt{\frac{D}{f}}$ is the dimensionless adsorption parameter representing the adsorption of the O form. The integration factor $(R_O)_m$ is defined as:

$$(R_O)_m = \exp\left(\frac{m\rho_O^2}{50}\right) \mathrm{erfc}\left(\rho_O\sqrt{\frac{m}{50}}\right) - \exp\left(\frac{(m-1)\rho_O^2}{50}\right) \mathrm{erfc}\left(\rho_O\sqrt{\frac{m-1}{50}}\right) . \qquad (2.168)$$

The response of a reversible reaction (2.146) depends on two dimensionless adsorption parameters, ρ_R and ρ_O. When $\rho_R = \rho_O$ the adsorbed species accomplish instantaneously a redox equilibrium after application of each potential pulse, thus no current remains to be sampled at the end of the potential pulses. The only current measured is due to the flux of the dissolved forms of both reactant and product of the reaction. For these reasons, the response of a reversible reaction of an adsorbed redox couple is identical to the response of the simple reaction of a dissolved redox couple (2.157), provided $\rho_R = \rho_O$. As a consequence, the real net peak current depends linearly on $\sqrt{f}$, and the peak potential is independent of the frequency. If the adsorption strength of the product decreases, i.e., the ratio $r = \frac{\beta_R}{\beta_O}$ increases, the net peak current starts to increase (Fig. 2.73). Under these conditions, the establishment of equilibrium between the adsorbed redox forms is prevented by the mass transfer of the product from the electrode surface. Thus, the redox reaction of adsorbed species contributes to the overall response, causing an increase of the current. In the limiting case, when $\beta_O \to 0$, the reaction (2.146) simplifies to reaction (2.144).

For the quasireversible case, (2.150), (2.153), (2.164) and (2.166) are combined with the following kinetic equation:

$$\frac{I}{nFA} = k_{\mathrm{sur}} \exp(\alpha_a \varphi)\left[\Gamma_R - \exp(-\varphi)\Gamma_O\right] \qquad (2.169)$$

The numerical solution is:

$$\Psi_m = \frac{\omega \exp(\alpha_a \varphi_m)\{X_m\}}{1 + \omega \exp(\alpha_a \varphi_m)\left[\frac{1}{\rho_R}\left(\frac{2}{\sqrt{50\pi}} + \frac{(R_R)_1}{\rho_R}\right) + \frac{\exp(-\varphi_m)}{\rho_O}\left(\frac{2}{\sqrt{50\pi}} + \frac{(R_O)_1}{\rho_O}\right)\right]} \qquad (2.170)$$

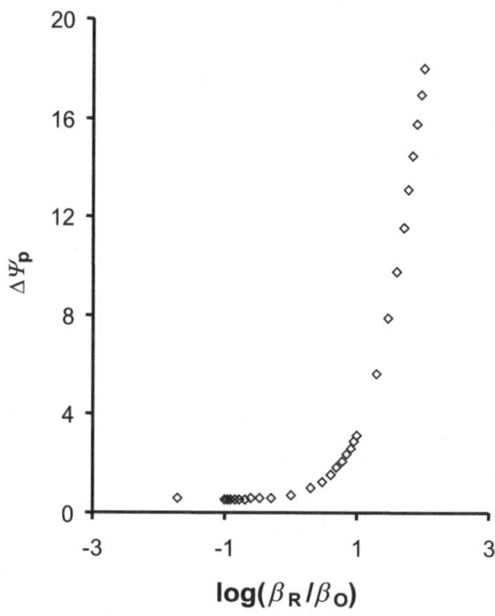

Fig. 2.73 Reversible reaction coupled with adsorption of both the reactant and product. Variation of $\Delta\Psi_p$ with the ratio $r = \frac{\beta_R}{\beta_O}$. For $r > 1$, $\beta_R = 0.1$ cm and $\beta_O = \frac{\beta_R}{r}$ and for $r < 1$, $\beta_O = 0.1$ cm and $\beta_R = r\beta_O$. The other conditions of the simulations are: $t_{\text{delay}} = 60$ s, $\Delta E = 10$ mV, $nE_{\text{sw}} = 50$ mV, $f = 10$ Hz, $D = 5 \times 10^{-6}$ cm^2 s^{-1}

where the term X_m is

$$X_m = \frac{1}{\rho_R}\left[1 - \exp\left(\frac{\rho_R^2 m}{50}\right)\text{erfc}\left(\rho_R\sqrt{\frac{m}{50}}\right) - \sum_{j=1}^{m-1}\Psi_j\left(\frac{2S_{m-j+1}}{\sqrt{50\pi}} + \frac{(R_R)_{m-j+1}}{\rho_R}\right)\right]$$
$$- \frac{\exp(-\varphi_m)}{\rho_O}\sum_{j=1}^{m-1}\Psi_j\left(\frac{2S_{m-j+1}}{\sqrt{50\pi}} + \frac{(R_O)_{m-j+1}}{\rho_O}\right) \tag{2.171}$$

The effect of the kinetic parameter ω on the dimensionless peak current for different values of the ratio r is shown in Fig. 2.74. The reaction is totally irreversible for $\log(\omega) < -2$ and electrochemically reversible for $\log(\omega) > 1$. The properties of a totally irreversible reaction are identical with those of reaction (2.144). Within the quasireversible region, a sharp quasireversible maximum is formed with a critical parameter $\omega_{\text{max}} = 0.912$, valid for the conditions given in Fig. 2.74. For $0.1 \leq r \leq 10$, the position of the quasireversible maximum is independent on r. Figure 2.75 shows the analysis performed by varying the frequency. The ordinate displays the ratio $\Delta\Psi_p f^{-0.5}$, which corresponds to the ratio $\Delta I_p f^{-1}$ in the real experiment. In the simulations, the standard rate constant was $k_{\text{sur}} = 113$ s^{-1}. The critical frequency associated with the maximum in Fig. 2.75 is $f_{\text{max}} = 120$ Hz. Hence, the standard rate constant calculated as $k_{\text{sur}} = \omega_{\text{max}} f_{\text{max}}$ yields $k_{\text{sur}} = 109.4$ s^{-1}, confirming the applicability of the procedure based on the quasireversible maximum for estimation of k_{sur} with a precision of about $\pm 10\%$. As previously mentioned for the pure surface electrode reaction (Sect. 2.5.1), the values of ω_{max} depend slightly

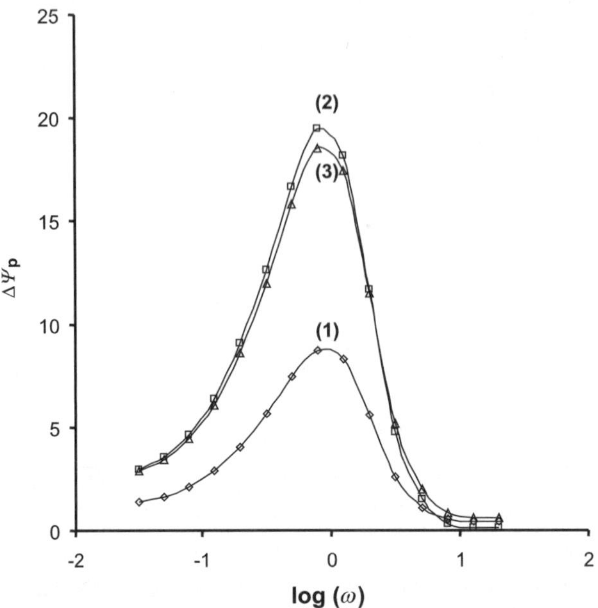

Fig. 2.74 Quasireversible reaction coupled with adsorption of both the reactant and product. Effect of the kinetic parameter ω on $\Delta\Psi_p$. Conditions of the simulations are: $t_{delay} = 60$ s, $\alpha_a = 0.5$, $\Delta E = 10$ mV, $nE_{sw} = 50$ mV, $f = 10$ Hz, $D = 5 \times 10^{-6}$ cm^2 s^{-1}, and $r = \frac{\beta_R}{\beta_O}$ is: 0.1 (*1*); 1 (*2*); and 10 (*3*)

on the electron transfer coefficient. If the transfer coefficient is not known, an average value $\omega_{avr} = 1.18 \pm 0.05$ can be used.

Similar to the pure surface electrode reaction, the response of reaction (2.146) is characterized by splitting of the net peak under appropriate conditions. The splitting occurs for an electrochemically quasireversible reaction and vanishes for the pure reversible reaction. Typical regions where the splitting emerges are $3 \leq \omega \leq 10$ and $0.1 \leq r \leq 10$ for $\alpha_a = 0.5$ and $nE_{sw} = 50$ mV. Contrary to the surface electrode reaction where the ratio of the split peak currents is solely sensitive to α_a, in the present system this ratio depends additionally on r. For instance, if $\alpha_a = 0.5$ and $r = 1$ the ratio is $\frac{\Psi_{p,a}}{\Psi_{p,c}} = 1$; for $r = 10$, $\frac{\Psi_{p,a}}{\Psi_{p,c}} > 1$; and $r = 0.1$, $\frac{\Psi_{p,a}}{\Psi_{p,c}} < 1$. Finally it is worth mentioning when experimentally possible, the electrode mechanism represented by (2.145) to (2.147) has to be simplified to a simple surface reaction (Sect. 2.5.1) in order to avoid the complexity arising from the effect of diffusion mass transport.

There are numerous analytically oriented studies developed upon adsorption coupled electrode reactions (2.144) and (2.146), which are summarized in the Sect. 3.1. For the purpose of verification of the theory, electrode mechanisms including reductions of a series of metallic ions in the presence of anion-induced adsorption [110], as well as electrode mechanisms at a mercury electrode of methylene blue [92], azobenzene [79], midazolam [115], berberine [111], jatrorubine [121], Cu(II)-sulfoxine and ferron complexes [122], Cd(II)- and Cu(II)-8-hydoxy-quinoline

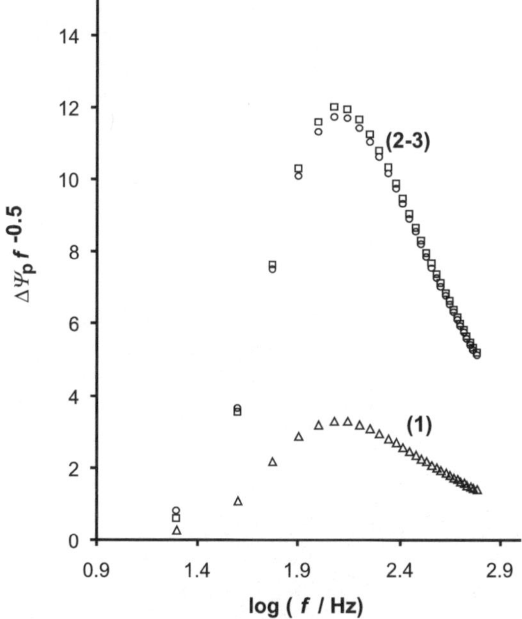

Fig. 2.75 Quasireversible reaction coupled with adsorption of both the reactant and product. Effect of f on the ratio $\Delta\Psi_p f^{-0.5}$. Conditions of the simulations are: $t_{\text{delay}} = 300$ s, $k_{\text{sur}} = 113$ s^{-1}, $\alpha_a = 0.5$, $\Delta E = 5$ mV, $nE_{\text{sw}} = 50$ mV, $D = 5 \times 10^{-6}$ cm^2 s^{-1}, and the ratio $r = \frac{\beta_R}{\beta_O}$ is: 0.1 (*1*); 1 (*2*); and 10 (*3*)

complexes [123], Cd(II) complex with ferron (8-Hydroxy-7-iodo-5-quinolinesulfonic acid) [119], have been used as model systems. In another set of studies the theory has been utilized to characterize the adsorption complicated electrode reactions of altertoxin I [103], DNA [124], Mo(VI)-fulvic acid complex [95], Cu(II) complexes with saccharin and cysteine [125], antidepressant drug fluoxetine [126], and a series of amalgam forming metals in the presence of surface active substances [127]. effect of the electrode kinetics in the case of reaction (2.144) has been extensively studied by comparing responses of a series of metallic ions adsorbed on the mercury electrode in the form of complexes formed with anions of the supporting electrolyte, such as Cl^-, ClO_4^- and NO_3^- ions (Fig. 2.76). The experimentally observed dependencies correspond to the effect of the electrochemical reversibility on the normalized net peak current of the electrode reaction (compare Fig. 2.72 with Fig. 2.76). The appearance of a maximum on the experimental curves in Fig. 2.76 confirms that the reduction pathway of metal cations involves adsorption of the complex formed with the present anions. The strongest anion-induced adsorption was observed for Bi^{3+} in a chloride containing medium. The reduction net peak current of Bi^{3+} depends parabolically on $\log c(Cl^-)$, which could be explained by the influence of the chloride ions on the charge transfer kinetics of the electrode reaction.

The theory for totally irreversible reaction of an adsorbed reactant was tested by the experiments with the organic compounds berberine [111] and midazolam [115].

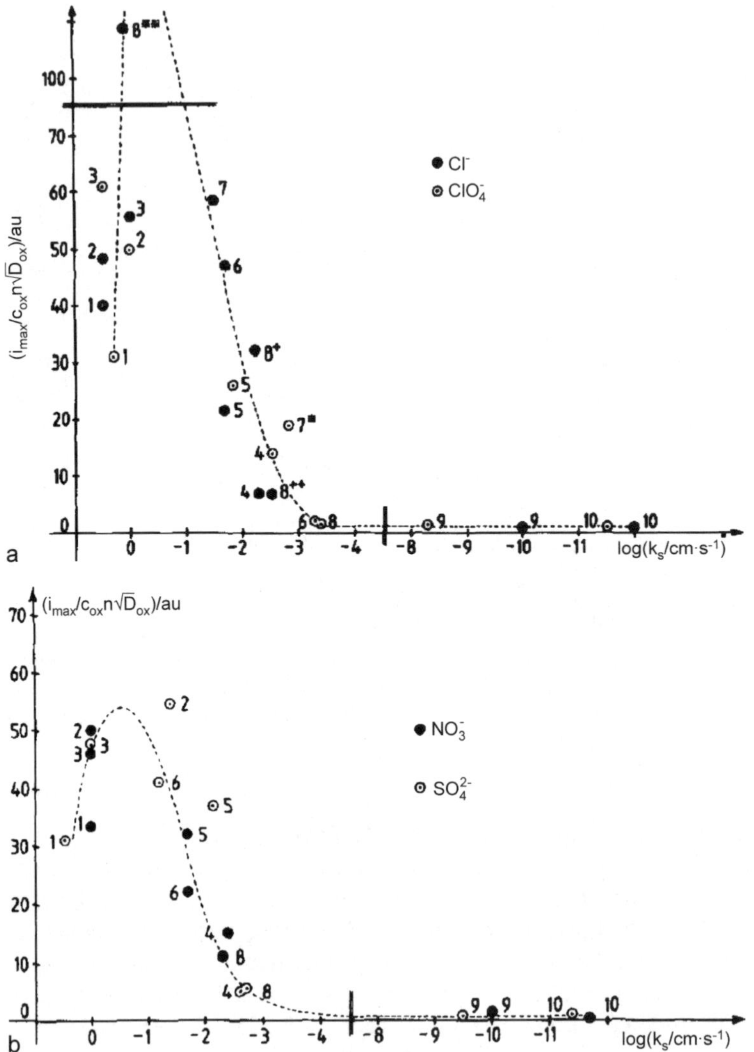

Fig. 2.76a,b Dependence of the normalized net peak currents (in arbitrary units) of Tl^{1+} (*1*); Cd^{2+} (*2*); Pb^{2+} (*3*); Zn^{2+} (*4*); UO_2^{2+} (*5*); Cu^{2+} (*6*); In^{3+} (*7*); Bi^{3+} (*8*); Ni^{2+} (*9*) and Co^{2+} (*10*) on the logarithm of the standard rate constant. The anions of the supporting electrolytes are: **a** Cl^- ($\bullet$) and ClO_4^- ($\odot$) and **b** NO_3^- ($\bullet$) and SO_4^{2-} ($\odot$). The other conditions are: $f = 225$ Hz, $E_{sw} = 32$ mV, $t_{delay} = 4$ s (reprinted from [110] with permission)

Berberine is an alkaloid undergoing an irreversible four-electron and three-proton reduction to the electrochemically inactive compound canadine, which is also adsorbed on the mercury electrode surface. As predicted by the theory, the net peak current of berberine is a linear function of the frequency, whereas the peak current shifts linearly with $\log(f)$ with a slope of -45 mV. Based on the theoretically predicted value for the half-peak width, $\Delta E_{p/2} = (63.5 \pm 0.5)/\alpha_c n$ mV, the cathodic electron transfer coefficient for berberine was estimates as $\alpha_c = 0.32$. O'Dea et al. analyzed the totally irreversible reduction of the drug midazolam (8-chloro-6-(2-fluorophenyl)-1-methyl-4H-imidazo[1,5-a][1,4]benzodiazepine) [115]. These authors employed a sophisticated approach for estimation of the redox kinetic parameters based on a fitting of the experimental and theoretical voltammograms with the aid of COOL algorithm. In this study an important feature of the response of the totally irreversible reaction has been pointed out: For a totally irreversible reaction of an adsorbed reactant, the reverse component has the same sign as the forward one. As a consequence, the net peak is lower than the forward peak. For these reasons, for analytical purposes an amplitude of $nE_{sw} \geq 50\,\mathrm{mV}$ was recommended, which ensures a minimal reverse component and a maximal sensitivity of the net response.

The theory for the reaction of an adsorbed redox couple (2.146) has been exemplified by experiments with methylene blue [92], and azobenzene [79]. Both redox couples, methylene blue/leucomethylene, and azobenzene/hydrazobenzene adsorb strongly on the mercury electrode surface. The reduction of methlylene blue involves a very fast two-step redox reaction with a standard rate constants of $3000\,\mathrm{s}^{-1}$ and $6000\,\mathrm{s}^{-1}$ for the first and second step, respectively. Thus, for $f < 50\,\mathrm{Hz}$, the kinetic parameter for the first electron transfer is $\log(\omega) > 1.8$, implying that the reaction appears reversible. Therefore, regardless of the adsorptive accumulation, the net response of methylene blue is a small peak, the peak current of which depends linearly on $\sqrt{f}$. Increasing the frequency above $50\,\mathrm{Hz}$, the electrochemical

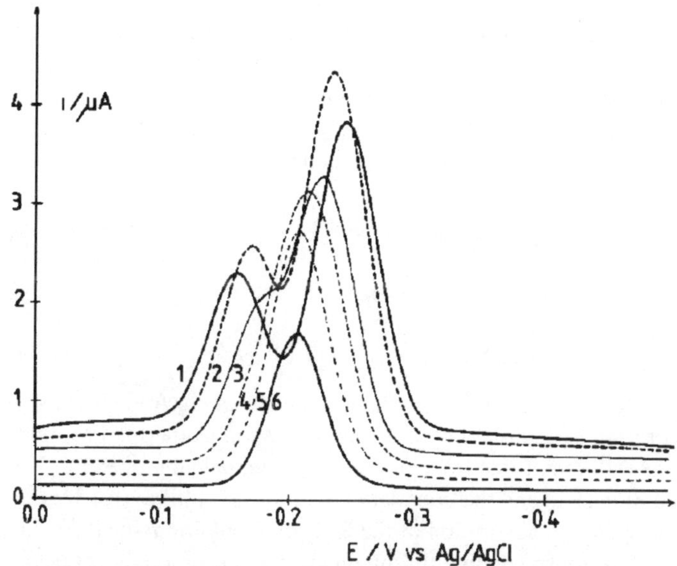

Fig. 2.77 The splitting of the net peaks of 5×10^{-6} mol/L methylene blue solution recorded in 0.9 mol/L NaNO$_3$ at pH $= 8$ under the influence of the SW amplitudes. The conditions are: $t_{acc} = 5$ s, $f = 100$ Hz, $\Delta E = 2$ mV, $E_{sw} = 120$ (*1*); 100 (*2*); 80 (*3*); 60 (*4*); 40 (*5*) and 20 mV (*6*) (reprinted from [92] with permission)

reversibility of the reaction decreases, causing an increase of the net peak current due to the effect of the quasireversible maximum shown in Fig. 2.75. Increasing the amplitude of the potential modulation, a splitting of the net peak of methylene blue was observed (Fig. 2.77), being in accord with the theoretical expectations for the reaction (2.146). The non equal heights of the split peaks indicate either $\alpha_c \neq 0.5$ or significant differences in the adsorption strength of methylene and leucomethylene on the mercury electrode surface.

The charge transfer kinetics of azobenzene at the mercury electrode is slower than that of methylene blue, thus the frequency interval provided by modern instrumentation ($10 < f/\text{Hz} < 2000$) allows variation of the electrochemical reversibility of the electrode reaction over a wide range [79]. The quasireversible maxima measured by the reduction of azobenzene in media at different pH are shown in Fig. 2.47 in the previous Sect. 2.5.1. The position of the quasireversible maximum depends on pH; hence the estimated standard rate constant obeys the following dependence: $k_{\text{sur}} = (62 - 12\,\text{pH})\,\text{s}^{-1}$ for pH ≤ 4. These results confirm the quasireversible maximum can be experimentally observed for a single electrode reaction by varying the frequency, as predicted by analysis in Fig. 2.75.

2.6.2 Electrode Reactions Coupled with Adsorption and Chemical Reactions

Electrode reactions coupled to adsorption equilibria and chemical reactions are among the most complex mechanisms treated in the theory of SWV. In the literature published so far, four types of adsorption coupled EC mechanisms have been considered [86, 128–130]. In all cases, the follow-up chemical reaction is an irreversible process (C_i). The simplest case is an EC$_i$ mechanism with adsorption of the reactant only [86]:

$$R_{(\text{ads})} \rightleftarrows R_{(\text{aq})} \tag{2.172}$$

$$R_{(\text{aq})} \rightleftarrows O_{(\text{aq})} + n\,\text{e}^- \tag{2.173}$$

$$O_{(\text{aq})} \xrightarrow{k} P_{(\text{aq})} \tag{2.174}$$

As in the previous chapter, the semi-infinite diffusion at a planar electrode is considered, where the adsorption is described by a linear adsorption isotherm. The modeling of reaction (2.173) does not require a particular mathematical procedure. The model comprises equation (1.2) and the boundary conditions (2.148) to (2.152) that describe the mass transport and adsorption of the R form. In addition, the diffusion of the O form, affected by an irreversible follow-up chemical reaction, is described by the following equation:

$$\frac{\partial c_O}{\partial t} = D\frac{\partial^2 c_O}{\partial x^2} - k c_O \tag{2.175}$$

which is a simplified form of (2.49). The above equation is solved under conditions (2.51) to (2.53), with the following solution:

$$(c_O)_{x=0} = \int_0^t \frac{I(\tau)}{nFA\sqrt{D}} \frac{e^{-k(t-\tau)}}{\sqrt{\pi(t-\tau)}} \, d\tau \qquad (2.176)$$

The voltammetric characteristics of reaction (2.173) are quite predictable and they are a combination of features of the electrode reaction with adsorption of the reactant (Sect. 2.6.1) and the simple EC_i mechanism of a dissolved redox couple (Sect. 2.4.2).

When both components of the redox couple adsorb on the electrode surface, the mechanism (2.172)–(2.174) transforms into the following square scheme [128]:

$$
\begin{array}{ccc}
& & k_v \\
R_{(aq)} & \leftrightarrows & O_{(aq)} + ne^- \to P \\
\updownarrow & & \updownarrow \\
R_{(ads)} & \leftrightarrows & O_{(ads)} + ne^-
\end{array}
\qquad (2.177)
$$

In the latter mechanism, only the dissolved form of O decays to the final electroinactive form P by an irreversible follow-up chemical reaction. That chemical reaction will be called a volume reaction, since it proceeds in the solution volume adjacent to the electrode surface, and it has the rate constant k_v (also called the volume rate constant). The diffusion of the O form is described by an equation equivalent to (2.175), which is solved under boundary conditions defined by (2.163) to (2.165). Details of the mathematical procedure are given in [128].

Assuming that both the adsorbed and the dissolved form of O participate in follow-up chemical reactions, the mechanism (2.177) transforms into the following general mechanism [129]:

$$
\begin{array}{ccc}
& & k_v \\
R_{(aq)} & \leftrightarrows & O_{(aq)} + ne^- \to P \\
\updownarrow & & \updownarrow \\
& & k_s \\
R_{(ads)} & \leftrightarrows & O_{(ads)} + ne^- \to P
\end{array}
\qquad (2.178)
$$

The dissolved form of O decays to the final electroinactive product via a volume chemical reaction occurring in the diffusion layer with the volume rate constant (k_v), whereas the adsorbed form participates in the surface chemical reaction confined to the electrode surface, characterized by a surface rate constant (k_s). These two chemical reactions proceed with different rates due to significant differences between the chemical nature of dissolved and adsorbed forms of O. Obviously, the mechanisms (2.172)–(2.174) and (2.177) are only limiting cases of the general mechanism (2.178).

For modeling of the mass transfer of the O form, besides differential equation (2.175) and the boundary conditions (2.163) and (2.165), the following condition is required, in order to take into account the effect of the surface chemical reaction:

$$t > 0, \quad x = 0: \frac{I}{nFA} = -D\left(\frac{\partial c_O}{\partial x}\right)_{x=0} + \frac{d\Gamma_O}{dt} - k_s\Gamma_O \qquad (2.179)$$

This equation shows that the current is produced by the flux of the O form and by the variation of its surface concentration. The last term of (2.179) shows that a part of the surface concentration is lost by the chemical reaction at a rate equal to the product of the surface concentration and the surface rate constant. Other details of the mathematical procedure can be found in [129]. The solution reads:

$$
\begin{aligned}
(c_O)_{x=0} = {} & \frac{a}{p-q} \int_0^t \frac{I(\tau)}{nFA\sqrt{D}} \left[p\exp(p(t-\tau)) - q\exp(q(t-\tau)) \right] d\tau \\
& - \frac{ak_s}{p-q} \int_0^t \frac{I(\tau)}{nFA\sqrt{D}} \left[\exp(p(t-\tau)) + \exp(q(t-\tau)) \right] d\tau \\
& - \frac{a^2}{p-q} \int_0^t \frac{I(\tau)}{nFA\sqrt{D}} \left[\sqrt{k_v+p}\,\mathrm{erf}\left(\sqrt{(k_v+p)(t-\tau)} \right) \right] d\tau \\
& + \frac{a^2}{p-q} \int_0^t \frac{I(\tau)}{nFA\sqrt{D}} \left[\sqrt{k_v+q}\,\mathrm{erf}\left(\sqrt{(k_v+q)(t-\tau)} \right) \right] d\tau \quad (2.180)
\end{aligned}
$$

where $p = \frac{B-\sqrt{B^2-4C}}{2}$ and $q = \frac{B+\sqrt{B^2-4C}}{2}$. Here $B = 2k_s + a^2$, $C = k_s^2 - a^2 k_v$, and $a = \frac{\sqrt{D}}{\beta}$, where β is the adsorption constant. The final integral equation describing a reversible electrode reaction is obtained by combining (2.180), (2.153), and the Nernst equation (1.8). The numerical solution is:

$$
\Psi_m = \frac{Q_m - X_m}{\frac{\rho}{p'-q'}\left[M_1 - \kappa_s MM_1 - \rho\left(RR_1 - RRR_1\right)\right] + \left(\frac{2}{\sqrt{50\pi}} + \frac{R_1}{\rho}\right) e^{\varphi_m}} \quad (2.181)
$$

where

$$
Q_m = \left[1 - e^{\frac{\rho^2 m}{50}} \mathrm{erfc}\left(\rho\sqrt{\frac{m}{50}}\right) - \sum_{j=1}^{m-1} \Psi_j \left(\frac{2S_{m-j+1}}{\sqrt{50\pi}} + \frac{R_{m-j+1}}{\rho}\right)\right] e^{\varphi_m} \quad (2.182)
$$

and

$$
\begin{aligned}
X_m = {} & \frac{\rho}{p'-q'} \left[\sum_{j=1}^{m-1} \Psi_j M_{m-j+1} - \kappa_s \sum_{j=1}^{m-1} \Psi_j MM_{m-j+1} \right. \\
& \left. - \rho\left(\sum_{j=1}^{m-1} \Psi_j RR_{m-j+1} - \sum_{j=1}^{m-1} \Psi_j RRR_{m-j+1} \right) \right] \quad (2.183)
\end{aligned}
$$

In the derivation of the numerical solution, it was assumed that the adsorption of both components of the redox couple is attributed with an identical adsorption constant. Here,

$$
p' = \kappa_s + \frac{\rho}{2} - \sqrt{\kappa_s\rho^2 + \frac{\rho^4}{4} - \kappa_v\rho^2}\,,
$$

$$
q' = \kappa_s + \frac{\rho}{2} + \sqrt{\kappa_s\rho^2 + \frac{\rho^4}{4} - \kappa_v\rho^2}\,, \qquad \rho = \frac{1}{\beta}\sqrt{\frac{D}{f}}
$$

is a dimensionless adsorption parameter, and $\kappa_s = \frac{k_s}{f}$ and $\kappa_v = \frac{k_v}{f}$ are dimensionless chemical rate parameters of the surface and volume follow-up chemical reaction, respectively. The numerical integration factor R_m is defined by (2.155). Other numerical integration factors are defined as follows:

$$M_m = \left(e^{\frac{p'}{50}m} - e^{\frac{p'}{50}(m-1)} \right) - \left(e^{\frac{q'}{50}m} - e^{\frac{q'}{50}(m-1)} \right) \tag{2.184}$$

$$MM_m = \frac{1}{p'} \left(e^{\frac{p'}{50}m} - e^{\frac{p'}{50}(m-1)} \right) + \frac{1}{q'} \left(e^{\frac{q'}{50}m} - e^{\frac{q'}{50}(m-1)} \right) \tag{2.185}$$

$$RR_m = \frac{1}{\sqrt{b\pi}} \left[\mathrm{erf}\left(\sqrt{\frac{bm}{50}} \right) \sqrt{\pi} \left(\frac{bm}{50} - \frac{1}{2} \right) + e^{-\frac{bm}{50}} \sqrt{\frac{bm}{50}} \right] -$$
$$\frac{1}{\sqrt{b\pi}} \left[\mathrm{erf}\left(\sqrt{\frac{b(m-1)}{50}} \right) \sqrt{\pi} \left(\frac{b(m-1)}{50} - \frac{1}{2} \right) + e^{-\frac{b(m-1)}{50}} \sqrt{\frac{b(m-1)}{50}} \right] \tag{2.186}$$

$$RRR_m = \frac{1}{\sqrt{c\pi}} \left[\mathrm{erf}\left(\sqrt{\frac{cm}{50}} \right) \sqrt{\pi} \left(\frac{cm}{50} - \frac{1}{2} \right) + e^{-\frac{cm}{50}} \sqrt{\frac{cm}{50}} \right] -$$
$$-\frac{1}{\sqrt{c\pi}} \left[\mathrm{erf}\left(\sqrt{\frac{c(m-1)}{50}} \right) \sqrt{\pi} \left(\frac{c(m-1)}{50} - \frac{1}{2} \right) + e^{-\frac{c(m-1)}{50}} \sqrt{\frac{c(m-1)}{50}} \right] \tag{2.187}$$

where $b = \kappa_v + p'$ and $c = \kappa_v + q'$.

Voltammetric features of adsorption coupled EC_i mechanisms (2.177) [128] and (2.178) [129] are rather unpredictable and deviate strongly from the EC_i mechanism of a dissolved redox couple. Their voltammetric behaviour is mainly controlled by the adsorption parameter ρ, and the dimensionless chemical parameters $\kappa_s = \frac{k_s}{f}$ and $\kappa_v = \frac{k_v}{f}$ reflecting the influence of the kinetics of the surface and volume follow-up chemical reaction, respectively. Figure 2.78 depicts the variation of the dimensionless net peak current with κ_v, for a variety of adsorption strengths of the redox couple, obtained by simulations of the mechanism (2.177). Instead of decreasing, the $\Delta\Psi_p$ enlarges by accelerating the volume chemical reaction. The increase of $\Delta\Psi_p$ is a consequence of the enhancement of the reverse component of the response. Beside the peculiar variation of $\Delta\Psi_p$, the peak potential shifts in a positive direction with a slope of $\frac{\Delta E_p}{\Delta\log(\kappa_v)} = 2.303\frac{RT}{2nF}$. Accordingly, the overall voltammetric behavior is totally opposite compared to the EC_i mechanism of a dissolved redox couple. If the adsorption is very strong, the mechanism (2.177) turns into the simple surface EC_i mechanism (Sect. 2.5.3). Consequently, the $\Delta\Psi_p$ decreases by increasing the rate of the follow-up chemical reaction (curve 1 in Fig. 2.78).

By analysing the mechanism (2.178) it was found that the effect of the surface follow-up chemical reaction is even more sever than the volume one. For instance, for a moderate adsorption ($\beta = 0.01\,\mathrm{cm}$), the influence of the surface and volume chemical reaction is measurable for $\log(\kappa_s) \geq 10^{-9}$ and $\log(\kappa_v) \geq 5 \times 10^{-3}$, respectively. Beside the surface rate constant, the overall effect of the surface chemical

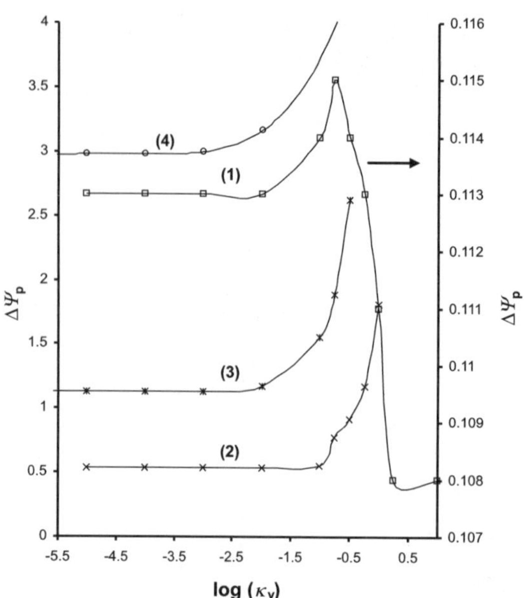

Fig. 2.78 Dependence of $\Delta\Psi_p$ on $\log(\kappa)$ for reaction scheme (2.177). The adsorption constant is $\beta = 0.1$ (*1*); 0.0033 (*2*); 0.002 (*3*) and 0.0012 cm (*4*). The other conditions are: $nE_{sw} = 50\,mV$, $\Delta E = 10\,mV$, $t_{delay} = 1\,s$, $E_{acc} = 0.10\,V$ vs. E^{θ} (with permission from [128])

reaction depends on the amount of the adsorbed material that is controlled by the adsorption constant and the duration of the accumulation period. For instance, for $\beta = 0.1\,cm$ and $t_{delay} = 100\,s$, the increase of κ_s from $\kappa_s = 10^{-6}$ to 10^{-3} causes a decrease of the reverse component of the response (Fig. 2.79), identical as for EC_i mechanism of a dissolved couple. However, further increase of the surface chemical reaction rate $(\kappa_s > 10^{-3})$ causes a strong increase of reverse component, followed by a significant shift toward more positive potentials.

Unique features of adsorption coupled EC_i mechanisms originate from the strong sensitivity of the response to the adsorption equilibria, as explained in the previous section. The adsorption equilibria are strongly affected by the follow-up chemical reactions causing irregular voltammetric features. Recall that for a simple EC_i mechanism of a dissolved redox couple, the follow-up chemical reaction consumes the electroactive material, which is always manifested as diminishing of both the reverse and the net component of the response. At the same time, as the electroactive product is lost by the chemical transformation, the response is shifted toward less potentials (in absolute values). For the adsorption coupled EC_i mechanisms (2.177) and (2.178), the follow-up chemical reactions also consume the electroactive material, but more importantly, they strongly affect the adsorption equilibria at the electrode surface. By acting of either surface or volume follow-up chemical reaction, the adsorption of the O form appears weaker, allowing the adsorption of the reactant R

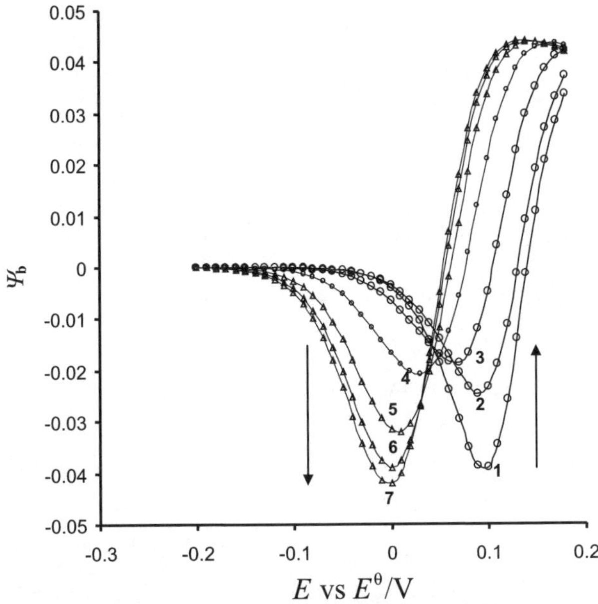

Fig. 2.79 The effect of the surface chemical parameter κ_s on the reverse component of the response for mechanism (2.178). The surface chemical parameter is $\log(\kappa_s) = -6$ (*1*); -5.4 (*2*); -4.8 (*3*); -3.9 (*4*); -3 (*5*); -2.7 (*6*) and -2.4 (*7*). The other conditions are: $\beta = 0.1$ cm, $k_v = 0$; $nE_{sw} = 50$ mV, $\Delta E = 10$ mV, $t_{delay} = 1$ s, $E_{acc} = 0.1$ V vs. E^{θ} (with permission from [129])

to prevail. As explained in the previous section, when the adsorption of the reactant R is becoming stronger than O, the dimensionless current strongly increases, whereas the oxidation appears more difficult, causing the peak to shift toward more positive potentials (see Fig. 2.73). This is one of the main reasons for the atypical behavior of the adsorption coupled EC$_i$ mechanisms.

If the follow-up chemical reactions regenerate the initial electroactive reactant, the mechanism (2.178) is transformed into a regenerative catalytic mechanism as represented by the following scheme [130]:

$$
\begin{array}{ccc}
& \xrightarrow{\;\;k_v\;\;} & \\
R_{(aq)} & \rightleftarrows & O_{(aq)} + ne^- \\
\uparrow\downarrow & & \uparrow\downarrow \\
R_{(ads)} & \rightleftarrows & O_{(ads)} + ne^- \\
& \xleftarrow{\;\;k_s\;\;} &
\end{array}
\tag{2.188}
$$

This mechanism is of particular significance for electroanalytical methods utilizing both adsorptive accumulation and catalytic regeneration for amplifying the analytical sensitivity. In the modeling of the mass transport of the O form, the equivalent procedure as described for the mechanism (2.178) is required. The mass transport of the R form is described by the differential equation (2.189) and the boundary conditions (2.148) to (2.150):

$$\frac{\partial c_R}{\partial t} = D\frac{\partial^2 c_R}{\partial x^2} + k_v c_O \tag{2.189}$$

In addition, the effect of the surface catalytic reaction is taken into account by the following boundary condition:

$$t > 0, \quad x = 0: \frac{I}{nFA} = D\left(\frac{\partial c_R}{\partial x}\right)_{x=0} - \frac{d\Gamma_R}{dt} - k_s\Gamma_O \tag{2.190}$$

To solve (2.189), the following substitution is introduced:

$$\Phi = c_R + c_O \tag{2.191}$$

Deriving the latter equation with respect to the variable t, and considering (2.189) and (2.175), one obtains:

$$\frac{\partial \Phi}{\partial t} = D\frac{\partial^2 \Phi}{\partial x^2} \tag{2.192}$$

The latter equation is solved for the following initial and boundary conditions:

$$t = 0, \quad x \geq 0: \quad \Phi = c_R^*, \quad \Gamma_\Phi = 0 \tag{2.193}$$

$$t > 0, \quad x \to \infty: \quad \Phi \to c_R^* \tag{2.194}$$

$$x = 0: \quad D\left(\frac{\partial \Phi}{\partial x}\right)_{x=0} = \frac{d\Gamma_\Phi}{dt} \tag{2.195}$$

$$\beta\Phi_{x=0} = \Gamma_\Phi \tag{2.196}$$

where, $\Gamma_\Phi = \Gamma_R + \Gamma_O$.

Applying Laplace transform to (2.192) and the boundary conditions (2.193) to (2.196) one obtains the following solution:

$$\Phi_{x=0} = c_R^* \left(1 - \exp(a^2 t)\mathrm{erfc}(a\sqrt{t})\right) \tag{2.197}$$

where $a = \frac{\sqrt{D}}{\beta}$. The procedure for deriving (2.197) is similar to that used in the modeling of a simple electrode reaction coupled by the adsorption of the reactant only (Sect. 2.6.1). A combination of (2.191) and (2.197) yields:

$$(c_R)_{x=0} = c_R^* \left[1 - \exp(a^2 t)\,\mathrm{erfc}\left(a\sqrt{t}\right)\right] - (c_O)_{x=0} \tag{2.198}$$

Therefore, to obtain the final solution for the concentration of the R form at the electrode surface, one requires the solution for the O form, given by (2.180). Assuming an equal adsorption of both components of the redox couples, the numerical solution for a reversible electrode reaction is:

$$\Psi_m = \frac{Q_m - X_m}{\frac{\rho}{p'-q'}\left[M_1 - \kappa_s MM_1 - \rho\left(RR_1 - RRR_1\right)\right]} \tag{2.199}$$

where

$$Q_m = \left[1 - e^{\frac{\rho^2 m}{50}}\mathrm{erfc}\left(\rho\sqrt{\frac{m}{50}}\right)\right]\frac{e^{\varphi_m}}{1 + e^{\varphi_m}} \tag{2.200}$$

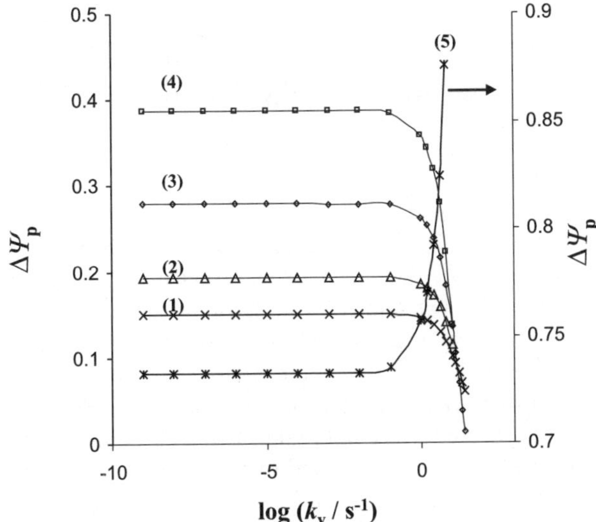

Fig. 2.80 Dependence of $\Delta\Psi_p$ on the $\log(k_v)$ for different adsorption strengths of the redox couple for the catalytic reaction (2.188). The adsorption constant is: $\beta = 0.033$ (*1*); 0.02 (*2*); 0.01 (*3*) and 0.005 cm (*4*). *Curve 5* (*right ordinate*) refers to the catalytic mechanism in the absence of adsorption (Chap. 2.4.4). The other conditions are: $nE_{sw} = 50$ mV, $\Delta E = 5$ mV, $\log(k_s/s^{-1}) = -8$, $f = 10$ Hz (with permission from [130])

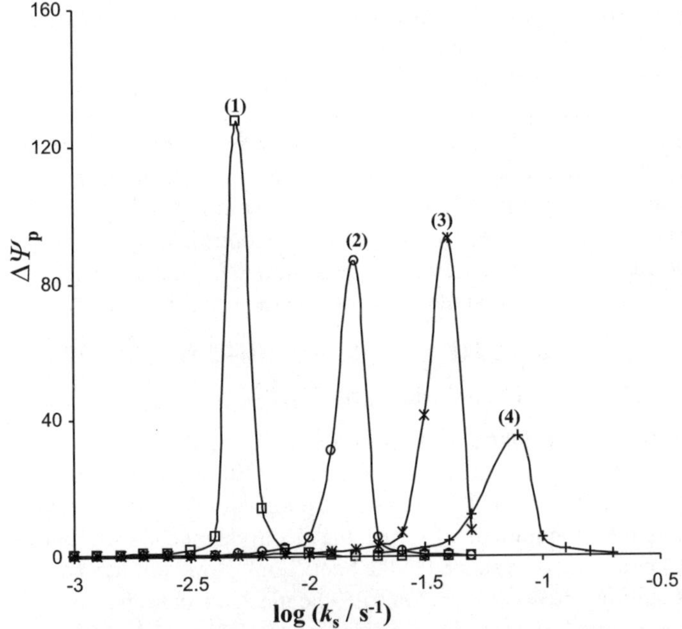

Fig. 2.81 Dependence of $\Delta\Psi_p$ on $\log(k_s)$ for different adsorption strength of the redox couple for catalytic reaction (2.188). The adsorption constant is: $\beta = 0.033$ (*1*); 0.02 (*2*); 0.0125 (*3*) and 0.009 cm (*4*). The other conditions are: $nE_{sw} = 50$ mV, $\Delta E = 10$ mV, $t_{delay} = 10$ s, $E_{acc} = 0.15$ V vs. E^θ, $\log(k_v/s^{-1}) = -8$, $f = 10$ Hz (with permission from [130])

and

$$X_m = \frac{\rho}{p' - q'} \left[\sum_{j=1}^{m-1} \Psi_j M_{m-j+1} - \kappa_s \sum_{j=1}^{m-1} \Psi_j M M_{m-j+1} \right.$$
$$\left. - \rho \left(\sum_{j=1}^{m-1} \Psi_j R R_{m-j+1} - \sum_{j=1}^{m-1} \Psi_j R R R_{m-j+1} \right) \right] \qquad (2.201)$$

All parameters of the above equations have the same meaning as for (2.181).

The effect of the volume and the surface catalytic reaction is sketched in Figs. 2.80 and 2.81, respectively. Obviously, the voltammetric behavior of the mechanism (2.188) is substantially different compared to the simple catalytic reaction described in Sect. 2.4.4. In the current mechanism, the effect of the volume catalytic reaction is remarkably different to the surface catalytic reaction, revealing that SWV can discriminate between the volume and the surface follow-up chemical reactions. The extremely high maxima shown in Fig. 2.81 correspond to the exhaustive reuse of the electroactive material adsorbed on the electrode surface, as a consequence of the synchronization of the surface catalytic reaction rate, adsorption equilibria, mass transfer rate of the electroactive species, and duration of the SW potential pulses. These results clearly reveal how powerful square-wave voltammetry is for analytical purposes when a moderate adsorption is combined with a catalytic regeneration of the electroactive material. This is also illustrated by a comparative analysis of the mechanism (2.188) with the simple surface catalytic reaction (Sect. 2.5.3) and the simple catalytic reaction of a dissolved redox couple (Sect. 2.4.4), given in Fig. 2.82.

All EC_i adsorption coupled mechanisms have been verified by experiments with azobenzene/hydrazobenzene redox couple at a hanging mercury drop electrode [86, 128, 130]. As mentioned in Sect. 2.5.3, azobenzene undergoes a two-electron and two-proton chemically reversible reduction to hydrazobenzene (reaction 2.202). In an acidic medium, hydrazobenzene rearranges to electrochemically inactive benzidine, through a chemically irreversible follow-up chemical reaction (reaction 2.203). The rate of benzidine rearrangement is controlled by the proton concentration in the electrolyte solution. Both azobenzene and hydrazobenzene, and probably benzidine, adsorb strongly on the mercury electrode surface.

$$C_6H_5-N=N-C_6H_{5(ads)} + 2\,e^- + 2\,H^+ \rightleftarrows C_6H_5-NH-NH-C_6H_{5(ads)} \qquad (2.202)$$
$$C_6H_5-HN-NH-C_6H_{5(ads)} \rightarrow H_2N-C_6H_4-C_6H_4-N_2H_{(ads)} \qquad (2.203)$$

The degree of adsorption can be controlled to some extent by addition of an organic solvent to the aqueous electrolyte, e.g., acetonitrile [128, 130] because an increasingly hydrophobic solvent mixture will shift the adsorption equilibrium to the solution side. In a pure aqueous medium of a low pH, the electrode mechanism follows a simple surface EC_i reaction, as explained in the Sect. 2.5.3. However, in an acidic aqueous medium containing 50% (v/v) acetonitrile, the mechanism transforms into one of the adsorption coupled EC_i reaction mechanisms (2.177) or (2.178). In such medium, the response increases in proportion to the rate of the follow-up chemical reactions, as evidenced by voltammograms depicted in Fig. 2.83. In Fig. 2.85,

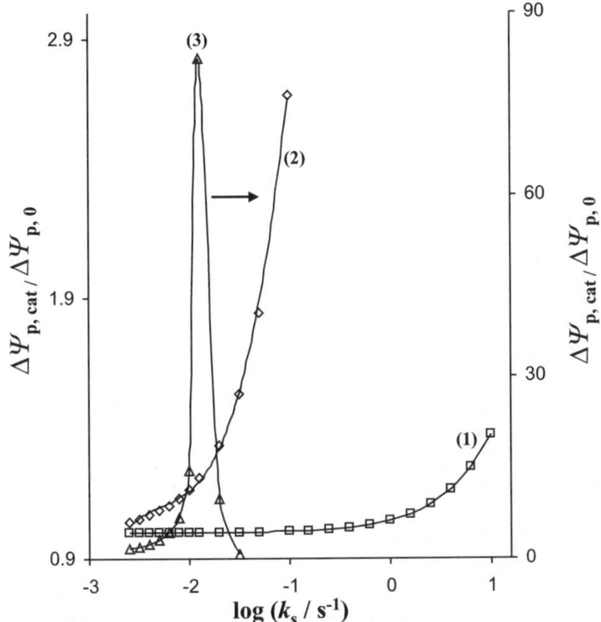

Fig. 2.82 Comparison of the effect of the catalytic reaction on the ratio $\Delta\Psi_{p,cat}/\Psi_{p,0}$ for a catalytic mechanism of a dissolved redox couple (*curve 1*, Sect. 2.4.4), surface catalytic mechanism (*curve 2*, Sect. 2.5.3.), and adsorption coupled catalytic reaction (*curve 3*, reaction (2.188), *right ordinate*). The simulation conditions are: $nE_{sw} = 10$ mV, $\Delta E = 10$ mV, $\beta = 0.02$ cm, $\log(k_v/s^{-1}) = -8$ (with permission from [130])

a comparison is given of the alteration of the net peak current with the rate of the follow-up chemical reaction for experiments performed in a pure aqueous medium (curve 1), and in a mixture of water and acetonitrile (curve 2). In a pure aqueous medium, when the electrode reaction follows the simple surface EC_i mechanisms the peak current decreases by enhancing the rate of the follow-up chemical reaction. On the contrary, in a water-acetonitrile mixture the electrode reaction is coupled with adsorption equilibria, thus the peak current enhances by accelerating the follow-up chemical reactions, in agreement with the theoretical predictions for reaction mechanisms (2.177) and (2.178).

In the presence of an oxidant, e.g., chlorate or bromate ions, the electrode reaction is transposed into an adsorption coupled regenerative catalytic mechanism. Figure 2.85 depicts the dependence of the azobenzene net peak current with the concentration of the chlorate ions used as an oxidant. Different curves in Fig. 2.85 correspond to different adsorption strength of the redox couple that is controlled by the content of acetonitrile in the aqueous electrolyte. In most of the cases, parabolic curves have been obtained, in agreement with the theoretically predicted effect for the surface catalytic reaction shown in Fig. 2.81. In a medium containing 50% (v/v) acetonitrile (curve 5 in Fig. 2.85) the current dramatically increases, confirming that moderate adsorption provides the best conditions for analytical application.

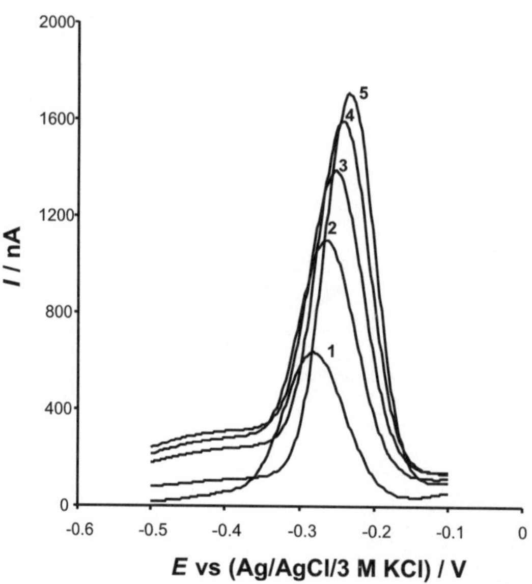

Fig. 2.83 Effect of the concentration of the acetic acid on the net peaks of 1×10^{-6} mol/L azobenezene solution recorded in sodium acetate solution containing 50% (v/v) acetonitrile and 0.0 (*1*); 0.175 (*2*); 0.35 (*3*); 0.70 (*4*) and 0.79 mol/L (*5*) acetic acid. The other experimental conditions were: $E_{sw} = 20$ mV, $\Delta E = 4$ mV, $f = 40$ Hz, $t_{delay} = 15$ s, $E_{acc} = -0.10$ V (with permission from [128])

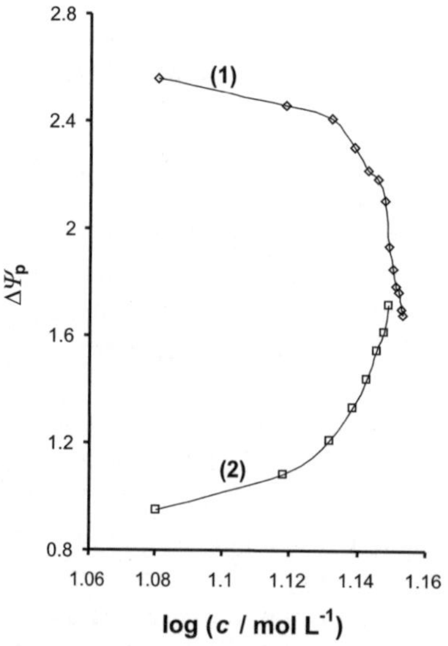

Fig. 2.84 Effect of the concentration of the acetic acid on the net peak currents of 1×10^{-6} mol/L azobenezene solution recorded in an acetate buffer (*1*) and an acetate buffer containing 50% (v/v) acetonitrile (*2*). The other experimental conditions are: $E_{sw} = 20$ mV, $\Delta E = 4$ mV, $f = 40$ Hz, $t_{delay} = 15$ s, $E_{acc} = 0.0$ V

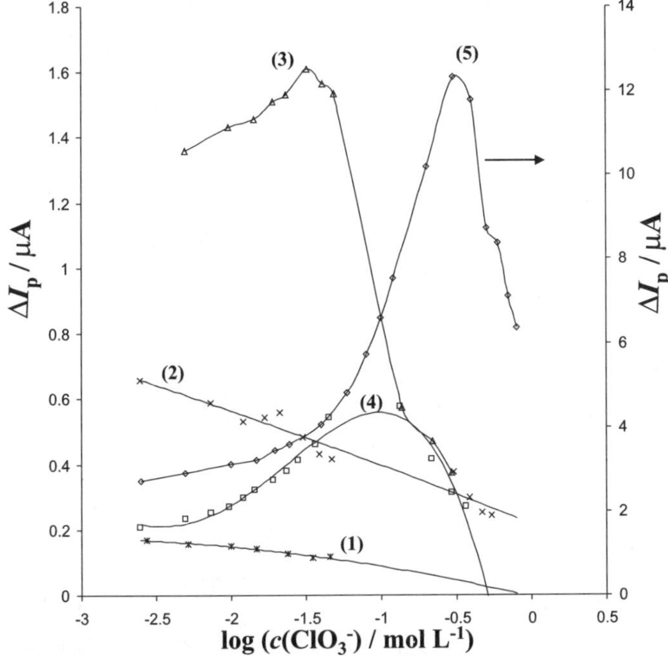

Fig. 2.85 The effect of the concentration of chlorate ions on the net peak current of 1×10^{-6} mol/l azobenzene solution recorded in an acetate buffer at pH = 4.75 containing 0 (*1*); 11 (*2*); 25 (*3*); 40 (*4*) and 50% (*5, right ordinate*) acetonitrile. The other experimental conditions are: $t_{delay} = 15$ s, $E_{acc} = -0.10$ V, $E_{sw} = 20$ mV, $\Delta E = 0.1$ mV, and $f = 60$ Hz. The accumulation is performed with stirring for *curves* (*2*), (*3*), and (*4*) (with permission from [130])

2.6.3 Electrode Reactions of Insoluble Salts

In this chapter electrode mechanisms based on the cathodic dissolution of a sparingly soluble salt deposited on a mercury or silver electrode are considered [131]. Miscellaneous inorganic anions such as halides, sulphide, disulphide, cyanide, phosphate, selenide, thiosulphite and sulphite, as well as a large group of sulphur containing organic compounds, porphyrins and flavins, although electrochemically inactive on a mercury or silver electrode, undergo chemical reaction with the electrode material to form a sparingly soluble salt [132]. To promote the salt precipitation, the mercury or silver electrode is anodically polarized in the course of the accumulation period in order to generate metal ions followed by formation of an insoluble compound with the analyte. As a consequence, a film of an insoluble compound is deposited on the electrode surface. During the cathodic potential scan, the film is stripped off the electrode surface yielding a cathodic stripping voltammetric response. Hence, the electrode reactions of this type are termed as "cathodic stripping reactions". This electrode mechanism attracted considerable interest in the theory of SWV due to its intriguing kinetic properties as well as its importance in electroanalysis [133, 138].

Most frequently, this electrode mechanism has been analyzed in connection with a mercury electrode; hence the following reaction schemes are pertinent to this electrode. Note, that the electrosorption mechanism can serve as a theoretical basis for these processes as well [135, 139, 140]. The simplest case of an accumulation/stripping equilibrium is given by the following equation:

$$Hg_{(l)} + L^{2-}_{(aq)} \rightleftarrows HgL_{(s)} + 2e^- \tag{2.204}$$

where L^{2-} is the symbol for a dissolved divalent ligand. With respect to the reacting ligand, reaction (2.204) is designated as a first-order cathodic stripping reaction [133]. In the case of a monovalent ligand, the electrode reaction has to be written as follows [134]:

$$Hg_{(l)} + 2L^-_{(aq)} \rightleftarrows HgL_{2(s)} + 2e^- \tag{2.205}$$

Assuming that a mercurous salt is formed, reaction (2.205) can be rewritten as:

$$2Hg_{(l)} + 2L^-_{(aq)} \rightleftarrows Hg_2L_{2(s)} + 2e^- \tag{2.206}$$

Reactions (2.205) and (2.206) are called second-order cathodic stripping reactions [134]. If the reacting ligand has a tendency to adsorb on the electrode surface, the following mechanisms are encountered [136, 137]:

$$L^{2-}_{(aq)} \rightleftarrows L^{2-}_{(ads)} \tag{2.207}$$

$$L^{2-}_{(ads)} + Hg_{(l)} \rightleftarrows HgL_{(s)} + 2e^- \tag{2.208}$$

$$L^-_{(aq)} \rightleftarrows L^-_{(ads)} \tag{2.209}$$

$$2L^-_{(ads)} + Hg_{(l)} \rightleftarrows HgL_{2(s)} + 2e^- \tag{2.210}$$

Reaction (2.208) is a first-order cathodic stripping reaction with adsorption of the ligand [136], whereas reaction (2.210) is of second order [137]. Considering a mercurous salt formation, reaction (2.210) is written in the following form:

$$2L^-_{(ads)} + 2Hg_{(l)} = Hg_2L_{2(s)} + 2e^- \tag{2.211}$$

For all reactions, the mass transport regime is controlled by the diffusion of the reacting ligand only, as the mercury electrode serves as an inexhaustible source for mercury ions. Hence, with respect to the mathematical modeling, reactions (2.205) and (2.206) are identical. This also holds true for reactions (2.210) and (2.211). Furthermore, it is assumed that the electrode surface is covered by a sub-monomolecular film without interactions between the deposited particles. For reactions (2.207) and (2.209) the ligand adsorption obeys a linear adsorption isotherm. Assuming semi-infinite diffusion at a planar electrode, the general mathematical model is defined as follows:

$$\frac{\partial c_L}{\partial t} = D\frac{\partial^2 c_L}{\partial x^2} \tag{2.212}$$

$$t = 0, \ x \geq 0: \ c_L = c_L^*; \quad \Gamma_{HgL} = 0; \qquad \text{(2.213) for reactions (2.204) to (2.210)}$$

$$\Gamma_L = 0; \qquad \text{(2.214) for reactions (2.208) to (2.210)}$$

$$t > 0, \quad x \to \infty: c_L \to c_L^*; \qquad \text{(2.215) for reactions (2.204) to (2.210)}$$

$$t > 0, \quad x = 0: \quad \beta(c_L)_{x=0} = \Gamma_L; \qquad \text{(2.216) for reactions (2.208) to (2.210)}$$

$$D \left(\frac{\partial c_L}{\partial x} \right)_{x=0} = \frac{I}{2FA} \, ; \qquad \text{(2.217) for reactions (2.204) to (2.205)}$$

$$D \left(\frac{\partial c_L}{\partial x} \right)_{x=0} - \frac{d\Gamma_L}{dt} = \frac{I}{2FA} \qquad \text{(2.218) for reactions (2.208) to (2.210)}$$

$$\frac{d\Gamma_{HgL}}{dt} = \frac{I}{2FA} \, ; \qquad \text{(2.219) for reactions (2.204) and (2.208)}$$

$$\frac{d\Gamma_{HgL_2}}{dt} = \frac{I}{4FA} \qquad \text{(2.220) for reactions (2.205) and (2.210)}$$

The solution for (2.212) for mechanisms without adsorption of the reacting ligand can be derived as described in Sect. 1.2. The solution of (2.219) is equivalent to that given by (2.94) in Sect. 2.5.1. The solution of (2.220) is:

$$\Gamma_{HgL_2} = \frac{1}{4} \int_0^t \frac{I(\tau)}{FA} \, d\tau \qquad (2.221)$$

For a quasireversible electrode reaction, the kinetic equation for reaction (2.204) can be attributed with a standard rate constant expressed in units of either $\mathrm{cm\,s^{-1}}$, (2.222), or $\mathrm{s^{-1}}$, (2.223):

$$\frac{I}{2FA} = k_s \exp(\alpha_a \varphi) \left[(c_L)_{x=0} - \exp(-\varphi) \frac{\Gamma_{HgL}}{r_s} \right] \qquad (2.222)$$

$$\frac{I}{2FA} = k_{sur} \exp(\alpha_a \varphi) \left[r_s (c_L)_{x=0} - \exp(-\varphi) \Gamma_{HgL} \right] \qquad (2.223)$$

Here, $\varphi = \frac{2F}{RT} \left(E - E^\theta \right)$ is a dimensionless potential and $r_s = 1\,\mathrm{cm}$ is an auxiliary constant. Recall that k_s in units of $\mathrm{cm\ s^{-1}}$ is heterogeneous standard rate constant typical for all electrode processes of dissolved redox couples (Sect. 2.2 to 2.4), whereas the standard rate constant k_{sur} in units of $\mathrm{s^{-1}}$ is typical for surface electrode processes (Sect. 2.5). This results from the inherent nature of reaction (2.204) in which the reactant $HgL_{(s)}$ is present only immobilized on the electrode surface, whereas the product $L_{(aq)}^{2-}$ is dissolved in the solution. For these reasons the cathodic stripping reaction (2.204) is considered as an intermediate form between the electrode reaction of a dissolved redox couple and the genuine surface electrode reaction [135]. The same holds true for the cathodic stripping reaction of a second order (2.205). Using the standard rate constant in units of $\mathrm{cm\ s^{-1}}$, the kinetic equation for reaction (2.205) has the following form:

$$\frac{I}{2FA} = k_s \exp(\alpha_a \varphi) \left[\frac{(c_L)_{x=0}^2}{c_s} - \exp(-\varphi) \frac{\Gamma_{HgL2}}{r_s} \right] , \qquad (2.224)$$

where $c_s = 1\,\mathrm{mol\,cm^{-3}}$ is the standard concentration.

For the electrode reaction (2.208), in a general case, two reaction pathways are possible. One is the reaction between the insoluble salt and the dissolved form of the reacting ligand, and the other one is between the insoluble salt and the adsorbed ligand. These two electrode reactions are expected to be attributed with different rate constants. Moreover, the formal potential of these reactions is not the same, the difference being dependent on the adsorption constant of the ligand. The reaction

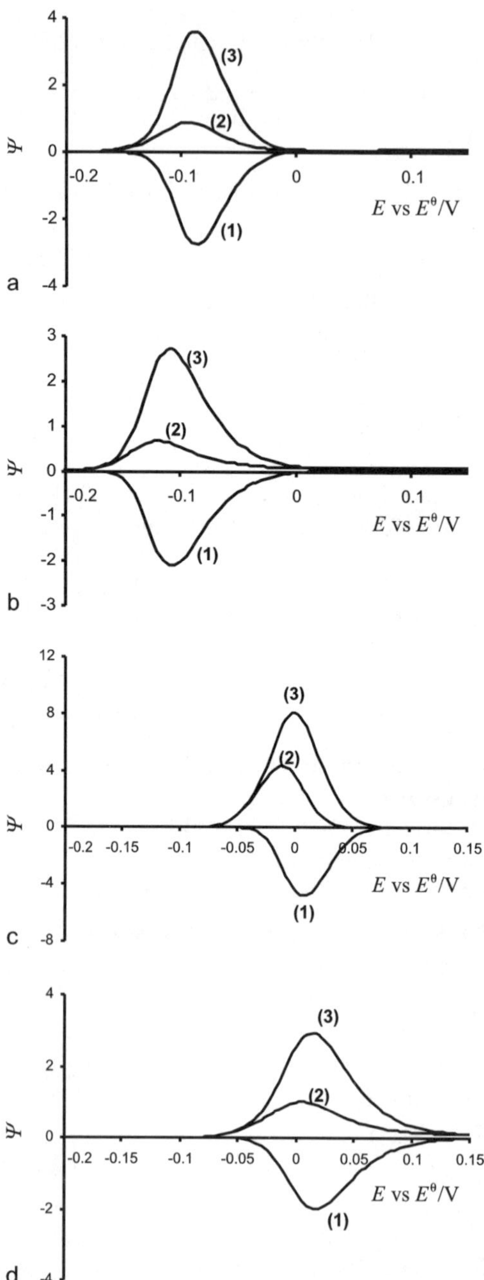

Fig. 2.86a–d Theoretical voltammograms simulated for the cathodic stripping mechanism (2.204) (**a**); (2.205) (**b**); (2.208) (**c**) and (2.210) (**d**). The conditions of simulations are: $k_s = 0.1\,\mathrm{cm\,s^{-1}}$ for a and b, and $k_{sur} = 10\,\mathrm{s^{-1}}$ for c and d. The other parameters are: $c = 1$ (for b), $\beta = 0.1\,\mathrm{cm}$ (for c), and $\rho = 0.01$ and $\chi = 100$ (for d). The other conditions are: $\alpha_a = 0.5$, $E_{sw} = 25\,\mathrm{mV}$, $\Delta E = 5\,\mathrm{mV}$, $t_{delay} = 5\,\mathrm{s}$, E_{acc} vs. $E^\theta = 0.15\,\mathrm{V}$. *Curves* (*1*), (*2*), and (*3*) correspond to the reduction, oxidation, and net component of the SW response

between the insoluble salt and the adsorbed ligand is of surface nature; hence it is expected to proceed faster than the other heterogeneous redox reaction. For these reasons we shall assume that reaction (2.208) proceeds as a genuine surface process associated with the following kinetic equation:

$$\frac{I}{2FA} = k_{\text{sur}} \exp(\alpha_a \varphi) \left[\Gamma_L - \exp(-\varphi)\Gamma_{\text{HgL}} \right] \tag{2.225}$$

Accordingly, the kinetic equation for reaction (2.210) is:

$$\frac{I}{2FA} = k_{\text{sur}} \exp(\alpha_a \varphi) \left[\frac{\Gamma_L^2}{\Gamma_s} - \exp(-\varphi)\Gamma_{\text{HgL}_2} \right] \tag{2.226}$$

where $\Gamma_s = 1 \, \text{mol cm}^{-2}$ is a standard surface concentration.

Substituting the solutions for surface concentrations into the corresponding kinetic equations, one obtains integral equations for each cathodic stripping reaction. Numerical solution for the quasireversible electrode mechanism (2.204) is:

$$\Psi_m = \frac{\kappa \, e^{\alpha_a \varphi_m} \left[1 - \frac{2}{\sqrt{50\pi}} \sum_{j=1}^{m-1} \Psi_j S_{m-j+1} - \frac{\gamma e^{-\varphi_m}}{50} \sum_{j=1}^{m-1} \Psi_j \right]}{1 + \kappa \, e^{\alpha_a \varphi_m} \left(\frac{2}{\sqrt{50\pi}} + \frac{\gamma e^{-\varphi_m}}{50} \right)} \tag{2.227}$$

Here $\Psi = \frac{I}{2FAc_L^* \sqrt{Df}}$ is dimensionless current, S_m is the integration factor defined as in (2.40), $\kappa = \frac{k_s}{\sqrt{Df}}$ is the common electrode kinetic parameter (Sect. 2.1.2) and $\gamma = \frac{1}{r_s}\sqrt{\frac{D}{f}}$ is a diffusion parameter. The numerical solutions for other cases can be derived according to the numerical procedure described in Sect. 1.2.

For each cathodic stripping mechanism, the dimensionless net peak current is proportional to the amount of the deposited salt, which is formed in the course of the deposition step. The amount of the salt is affected by the accumulation time, concentration of the reacting ligand, and accumulation potential. The amount of the deposited salt depends sigmoidally on the deposition potential, with a half-wave potential being sensitive to the accumulation time. If the accumulation potential is significantly more positive than the peak potential, the surface concentration of the insoluble salt is independent on the deposition potential. The formation of the salt is controlled by the diffusion of the ligand, thus the net peak current is proportional to the square root of the accumulation time. If reaction (2.204) is electrochemically reversible, the real net peak current depends linearly on the frequency, which is a common feature of all electrode mechanism of an immobilized reactant (Sect. 2.6.1). The net peak potential for a reversible reaction (2.204) is a linear function of the $\log(f)$ with a slope equal to $\frac{\Delta E_p}{\Delta \log(f)} = -2.303 \frac{RT}{2nF}$. A typical theoretical response of a quasireversible reaction (2.204) is depicted in Fig. 2.86a. The response is predominantly controlled by the kinetic parameter κ, and the diffusion parameter γ. The physical meaning of κ and γ is explained in the previous Sect. 2.1.2 and 2.6.1, respectively. The observed electrochemical reversibility is determined by the complex kinetic parameter $K = \kappa \sqrt{\gamma}$, or $K = \frac{k_s}{D^{1/4} f^{3/4} r_s^{1/2}}$. The reaction is quasireversible within the region $-1.5 \leq \log(K) \leq 1.5$. The intrinsic feature of the reaction (2.204) is the quasireversible maximum, the position of which depends on the electron transfer coefficient and the signal amplitude [135]. The interdependence between the critical kinetic parameter $K_{\max}$, associated with the position of the maximum, and the cathodic electron transfer coefficient α_c is $\log(K_{\max}) = -3.371\alpha_c + 1.805$ and $\log(K_{\max}) = -3.238\alpha_c + 1.899$ for amplitude of 25 and 10 mV, respectively [135]. According to these equations, $K_{\max}$ can be calculated for any values of the transfer coefficient. Amplitude larger than 50 mV is not recommended for kinetic measurements. If all the kinetic parameters are kept constant, the highest response is obtained for amplitude of about 30 mV. If $E_{sw} > 50$ mV, the half-peak width is markedly enlarged causing irregular peak shape.

Most of the voltammetric features of a reversible cathodic stripping reaction of a second order (2.205) are similar to reaction (2.204) [134]. The main differences arise due to the influence of the concentration of the ligand on the position of the voltammetric response. The peak potential depends linearly on $\log(c_L^*)$ with a slope of $\frac{\Delta E_p}{\log(c_L^*)} = -2.3\frac{RT}{2F}$, which is an inherent characteristic of a second-order reaction. Nevertheless, the dimensionless net peak current is virtually independent on c_L^*. Hence, the real net peak current is a linear function of the ligand concentration, which permits application of this mechanism for analytical purposes. A representative theoretical response of a quasireversible reaction (2.205) is shown in Fig. 2.86b. In addition to κ and γ, the dimensionless response is a function of the concentration parameter defined as $c = \frac{c_L^*}{c_S}$. The peak potential varies linearly with $\log(c_L^*)$ with a slope less than $-2.3RT/(2F)$, depending on the electrochemical reversibility of the reaction. For a given concentration of the ligand, the electrochemical reversibility is controlled by the kinetic parameter K, which is defined identically as for reaction (2.204). The position of the quasireversible maximum depends on c, besides the common dependencies of the maximum on α_c and E_{sw}. The interdependence between $K_{\max}$ and c obeys the following equation:

$$\log(K_{\max}) = -0.4589\log(c) + 0.1584 \tag{2.228}$$

The latter equation is valid for $E_{sw} = 20$ mV and $\alpha_c = 0.5$ [134]. Note that the sensitivity of the quasireversible maximum to the concentration of the ligand can serve as a qualitative diagnostic criterion for the reaction mechanism of a second order.

The voltammetric features of the adsorption coupled cathodic stripping reaction (2.208) are significantly different compared to previous reactions (2.204) and (2.205), due to the influence of the ligand adsorption [136]. The strength of adsorption affects both the position and the magnitude of the net peak. For a reversible reaction, the peak potential shifts linearly with $\log(\beta)$ with a slope equal to $\frac{\Delta E_p}{\log(\beta)} = 2.3\frac{RT}{2F}$. Particularly interesting is the effect of the adsorption constant on the net peak currents, shown in Fig. 2.87. A parabolic dependence is found, with a maximum located at $\log(\beta) = -2.4$, showing that moderate adsorption produces the highest peak current. A typical theoretical response of a quasireversible reac-

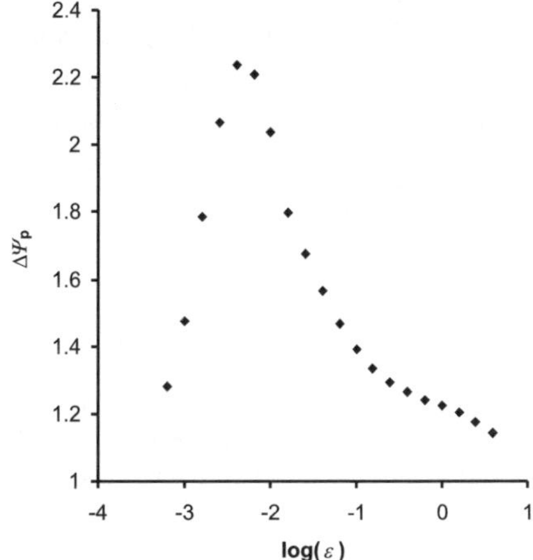

Fig. 2.87 Reversible electrode reaction (2.208). Dependence of $\Delta\Psi_p$ on the logarithm of the adsorption constant. Conditions of the simulations are: $D = 5 \times 10^{-6}$ cm^2 s^{-1}, $E_{sw} = 20$ mV, $f = 10$ Hz, $\Delta E = 10$ mV, $t_{delay} = 1$ s, $E_{acc} = 0.05$ V vs. E^{θ}

tion (2.208) is depicted in Fig. 2.86c. The narrow shape of the reverse (anodic) component is a consequence of the immobilization of the reacting ligand. In addition to κ and γ, the response is affected by the adsorption parameter $\rho = \frac{1}{\beta}\sqrt{\frac{D}{f}}$. For a given adsorption constant, the observed electrochemical reversibility depends on the kinetic parameter defined as $\omega = \lambda\gamma$, or $\omega = \frac{k_{sur}}{f}$. This reveals that the inherent properties of reaction (2.208) are very close to surface electrode reactions elaborated in Sect. 2.5. The quasireversible maximum is strongly pronounced, being represented by a sharp parabolic dependence of $\Delta\Psi_p$ vs. ω. The important feature of the maximum is its sensitivity to the adsorption constant, defined by the following equation:

$$\log(\omega_{max}) = 0.50\log(\beta) + 0.033 , \tag{2.229}$$

which holds for $\alpha_c = 0.5$ and $E_{sw} = 20$ mV. As typical for surface electrode processes, the net peak of the reaction (2.208) splits under appropriate conditions. For $E_{sw} = 50$ mV and $\omega > 0.6$, the splitting occurs over a narrow interval of the adsorption constant values, $0.01 < \beta/\text{cm} < 1$, which enables a rough estimation of the adsorption constant.

The second-order reaction with adsorption of the ligand (2.210) signifies the most complex cathodic stripping mechanism, which combines the voltammetric features of the reactions (2.205) and (2.208) [137]. For the electrochemically reversible case, the effect of the ligand concentration and its adsorption strength is identical as for reaction (2.205) and (2.208), respectively. A representative theoretical voltammogram of a quasireversible electrode reaction is shown in Fig. 2.86d. The dimensionless response is controlled by the electrode kinetic parameter ω, the adsorption parameter ρ, and the complex parameter $\chi = \frac{c_L^* \beta^2 \sqrt{f}}{\Gamma_s \sqrt{D}}$ that unifies the parameters representing the mass transport and the adsorption of the ligand. The variation of the net peak current with χ, calculated for different electrochemical reversibilities, is illustrated in Fig. 2.88. The net peak current depends parabolically on $\log(\chi)$, similar to the reaction (2.208). For a given values of χ and ρ, the observed electrochemical reversibility depends on the kinetic parameter ω. The, position of the quasireversible maximum depends on both the concentration and adsorption constant of the reacting ligand. The response of reaction (2.210) exhibits a splitting of the net peak. The effect of the complex parameter χ on the splitting is shown in Fig. 2.89. The potential separation between the split peaks depends on the concentration of the reacting ligand, which is a unique feature of this mechanism. This important characteristic can serve as a diagnostic criterion to distinguish reaction mechanism (2.208) and (2.210).

SWV has been applied to study electrode reactions of miscellaneous species capable to form insoluble salts with the mercury electrode such as: iodide [141, 142], dimethoate pesticide [143], sulphide [133, 144], arsenic [145, 146], cysteine [134, 147, 148], glutathione [149], ferron (7-iodo-8-hydroxyquinolin-5-sulphonic acid) [150], 6-propyl-2-thiouracil (PTU) [136], 5-fluorouracil (FU) [151], 5-azauracil (AU) [138], 2-thiouracil (TU) [138], xanthine and xanthosine [152], and selenium (IV) [153]. Verification of the theory has been performed by experiments at a mercury electrode with sulphide ions [133] and TU [138] for the simple first-order reaction, cystine [134] and AU [138] for the second-order reaction, FU for the first-order reaction with adsorption of the ligand [151], and PTU for the second-order reaction with adsorption of the ligand [137]. Figure 2.90 shows typical cathodic stripping voltammograms of TU and PTU on a mercury electrode. The order of the

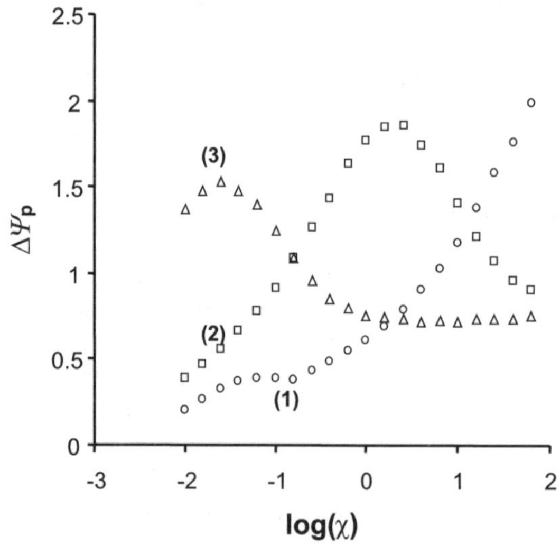

Fig. 2.88 Quasireversible electrode reaction (2.210). Dependence of $\Delta\Psi_p$ on the logarithm of the parameter χ for the electrode kinetic parameter $\omega = 0.1$ (*1*), 1 (*2*), and 10 (*3*). The other conditions of the simulations are: $\rho = 1$, $E_{sw} = 25$ mV, $\Delta E = 10$ mV, $t_{delay} = 0.1$ s, $E_{acc} = 0.15$ V vs. E^θ

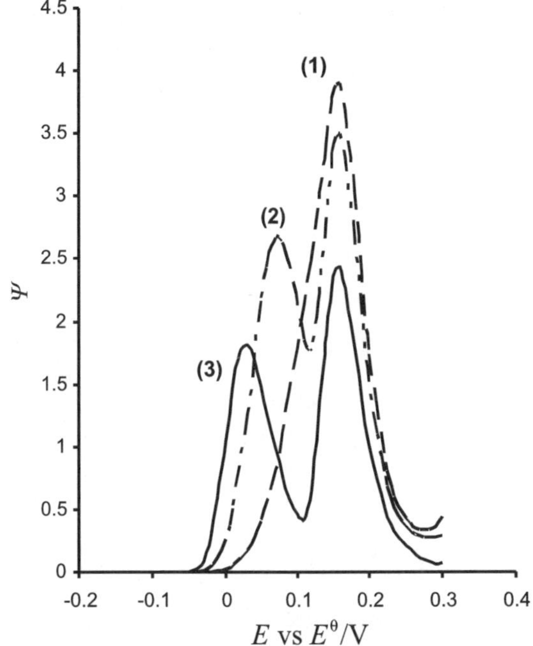

Fig. 2.89 Effect of the parameter χ on the splitting of the net SW voltammograms of reaction (2.210). The conditions of the simulations are: $\omega = 100$, $\rho = 0.001$, $E_{\mathrm{sw}} = 80\,\mathrm{mV}$, $\Delta E = 10\,\mathrm{mV}$, $t_{\mathrm{delay}} = 0.1\,\mathrm{s}$, $E_{\mathrm{acc}} = 0.30\,\mathrm{V}$ vs. E^{θ}, $\alpha_a = 0.5$

cathodic stripping reaction was established by analyzing the peak potential variation as a function of the concentration of the reacting compound. The adsorption of the reacting ligand was inferred from the shape of the oxidation component of the response as well as by inspecting the influence of the deposition performed at potential more negative than the peak potential. In the case of PTU, it was demonstrated that the adsorption strength varies depending on the amount of acetonitrile added to the electrolyte solution [136]. The effect of acetonitrile on the response parameters is given in Fig. 2.91. One part of the curve 1 is parabolic, which agrees with the theoretical predictions concerning the effect of adsorption for both reactions (2.208) and (2.210) (see Figs. 2.87 and 2.88). In addition, the peak potential shifts toward more negative values by increasing of the acetonitrile amount as a consequence of the declining of the adsorption strength of the ligand. Figure 2.92 shows the quasireversible maxima of AU measured for three different concentrations. As predicted by (2.228) the position of the quasireversible maximum is sensitive to the concentration of the reacting ligand for the reaction (2.205). This was the first experimental verification of this important feature of the second-order cathodic stripping reaction. As predicted by the theory, the cathodic stripping reaction coupled to adsorption of the reacting ligand is characterized by splitting of the net peak under appropriate experimental conditions. The splitting was observed for both FU [151] and PTU [136], confirming the adsorption of these compounds after cathodic disso-

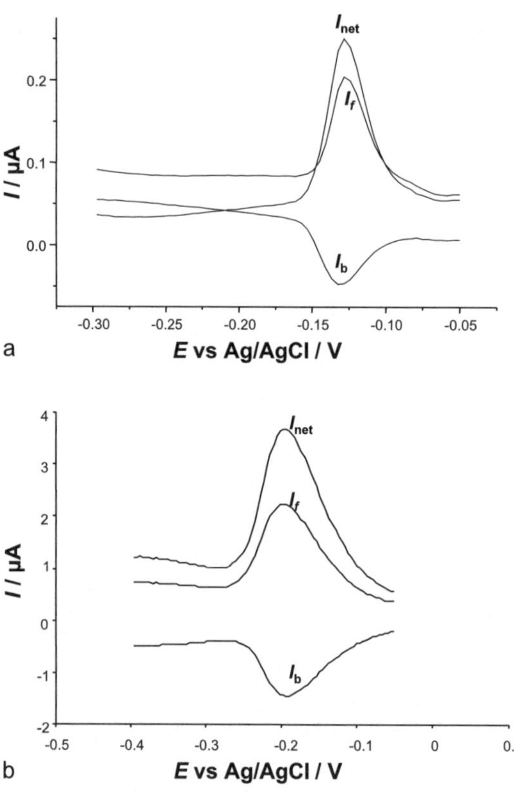

Fig. 2.90a,b a SW voltammogram of 1×10^{-5} mol/L 2-thiouracil recorded in 0.1 mol/L HCl solution. The other experimental conditions are: $E_{\mathrm{acc}} = 0$ V, $t_{\mathrm{delay}} = 45$ s, $E_{\mathrm{sw}} = 25$ mV, $\Delta E = 2$ mV, and $f = 150$ Hz. **b** SW voltammogram of 4×10^{-6} mol/L 6-propyl-2-thiouracil recorded in 1 mol/L KNO$_3$ solution. The other conditions are: $t_{\mathrm{delay}} = 30$ s, $E_{\mathrm{acc}} = -0.1$ V, $f = 150$ Hz, $E_{\mathrm{sw}} = 25$ mV, and $\Delta E = 4$ mV (with permission from [138])

lution of the corresponding mercury salt. Comparing the experimentally measured and theoretically calculated net SW voltammograms, the adsorption constant of FU and PTU have been determined to be about 0.1 cm, for both compounds [138].

2.7 Square-Wave Voltammetry Applied to Thin-Layer Cell

Square-wave voltammetry applied to experiments with a thin-layer electrochemical cell is a valuable analytical tool for determination of small amounts of analytes [46, 154–157], e.g., the determination of drugs and species with biological activity [158]. Over the past decades, SWV has been frequently applied to study physiologically active compounds embedded in a thin-film that is imposed on an electrode surface [78, 159]. Moreover, a graphite electrode modified with a thin-film

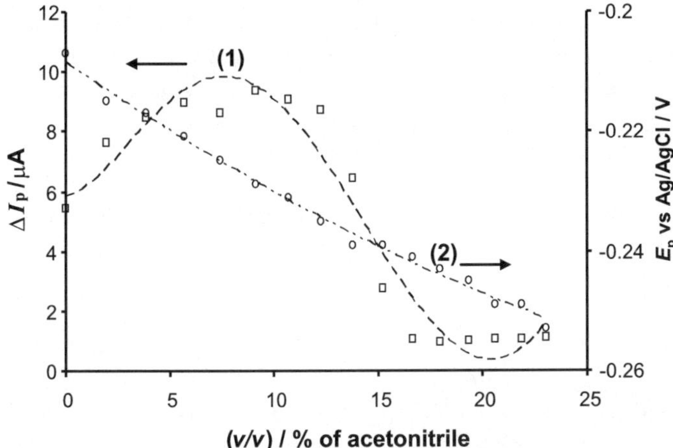

Fig. 2.91 The influence of acetonitrile on the net SW peak currents ΔI_p (*left axis, curve 1*) and peak potential E_p (*right axis, curve 2*) of 6-propyl-2-thiouracil (PTU). The experimental conditions are: $c(\text{PTU}) = 1 \times 10^{-4}$ mol/L, $f = 50$ Hz, $E_{sw} = 20$ mV, $\Delta E = 2$ mV, $t_{delay} = 10$ s, $E_{acc} = -0.10$ V (with permission from [136])

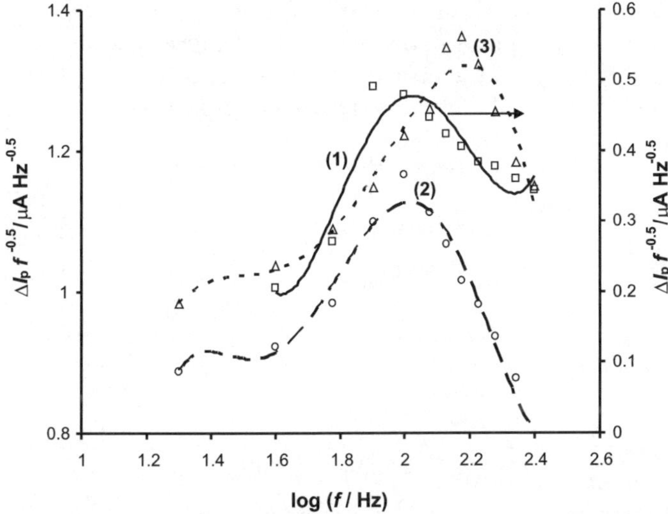

Fig. 2.92 Influence of the concentration of 5-azauracil (AU) on the position of the quasireversible maximum recorded in 1 mol/L KNO_3 solution. The concentrations of AU were: $c(\text{AU}) = 5 \times 10^{-6}$ (*1, right ordinate*), 5×10^{-5} (*2*), and 5×10^{-4} mol/L (*3*). The other conditions are: $t_{delay} = 30$ s, $E_{acc} = 0.30$ V, $E_{sw} = 25$ mV, and $\Delta E = 2$ mV (with permission from [138])

of a water immiscible organic solvent containing a lipophilic redox probe, became an alternative and powerful experimental tool to study charge transfer phenomena across an interface between two immiscible liquids [155, 160, 161].

The current chapter gives an account on the theory of SWV for systems occurring in a finite diffusion space. The experimental verification of the theoretical predictions are presented in Chap. 4, devoted to the application of SWV to study ion transfer reactions with the help of three-phase and thin-film electrodes. Note that the theory regarding electrode processes in thin mercury films is separately discussed in Chap. 2.3.

In the modeling of the electrode reaction (2.230) proceeding in a thin-layer cell or within a thin film on the electrode surface, the differential equation (1.2) and (1.3), together with the following boundary conditions (2.231) to (2.233) have to be considered:

$$R \rightleftarrows O + ne^- \tag{2.230}$$

$$t = 0, \quad 0 \le x \le L: \quad c_R = c_R^*, \quad c_O = 0 \tag{2.231}$$

$$t > 0, \quad x = 0: \quad D\left(\frac{\partial c_R}{\partial x}\right) = -D\left(\frac{\partial c_O}{\partial x}\right) = \frac{I}{nFA} \tag{2.232}$$

$$t > 0, \quad x = L: \quad D\left(\frac{\partial c_R}{\partial x}\right) = -D\left(\frac{\partial c_O}{\partial x}\right) = 0 \tag{2.233}$$

The physical meaning of condition (2.233) is that the diffusion of the electroactive species is blocked at the distance $x = L$, i.e., where L is the thickness of the film. This boundary condition complicates significantly the mathematical procedure compared to the semi-infinite diffusion case. To resolve the mathematical complexity, recently a novel mathematical approach has been developed which is based on the modification of the step function method [162], as elaborated in more detail in the Appendix. The numerical solution for a reversible electrode reaction is given by [155]:

$$\Psi_m = \frac{\frac{1}{\sqrt{f}} + [\exp(-\varphi_m) + 1] \sum_{j=1}^{m-1} \Psi_j S_{m-j+1}}{-S_1[\exp(-\varphi_m) + 1]} \tag{2.234}$$

Here, numerical integration factor is defined as:

$$S_m = \frac{1}{50pf} \sum_{i=p(m-1)}^{pm} \Phi_i \tag{2.235}$$

where

$$\Phi_i = \frac{\sqrt{\frac{50p}{\pi i}} + \sqrt{\frac{50p}{\pi i}} \exp\left(\frac{-\Lambda^2 50p}{i}\right) - \sum_{j=1}^{i-1} \Phi_i M_{i-j+1}}{M_1 - 1} \tag{2.236}$$

and

$$M_i = \mathrm{erfc}\left(\Lambda\sqrt{\frac{50p}{i+1}}\right) - \mathrm{erfc}\left(\Lambda\sqrt{\frac{50p}{i}}\right) \tag{2.237}$$

$\Lambda = L\sqrt{\frac{f}{D}}$ is the dimensionless thickness parameter, p is the number of time subintervals, and i is the series number of time subintervals (see Appendix).

The voltammetric features of a reversible reaction are mainly controlled by the thickness parameter $\Lambda = L\sqrt{\frac{f}{D}}$. The dimensionless net peak current depends sigmoidally on $\log(\Lambda)$, within the interval $-0.2 \le \log(\Lambda) \le 0.1$ the dimensionless net peak current increases linearly with Λ. For $\log(\Lambda) < -0.5$ the diffusion exhibits no effect to the response, and the behavior of the system is similar to the surface electrode reaction (Sect. 2.5.1), whereas for $\log(\Lambda) > 0.2$, the thickness of the layer is larger than the diffusion layer and the reaction occurs under semi-infinite diffusion conditions. In Fig. 2.93 is shown the typical voltammetric response of a reversible reaction in a film having a thickness parameter $\Lambda = 0.632$, which corresponds to $L = 2\,\mu\text{m}$, $f = 100\,\text{Hz}$, and $D = 1 \times 10^{-5}\,\text{cm}^2\,\text{s}^{-1}$. Both the forward and backward components of the response are bell-shaped curves. On the contrary, for a reversible reaction under semi-infinite diffusion condition, the current components have the common non-zero limiting current (see Figs. 2.1 and 2.5). Furthermore, the peak potentials as well as the absolute values of peak currents of both the forward and backward components are virtually identical. The relationship between the real net peak current and the frequency depends on the thickness of the film. For $L > 10\,\mu\text{m}$ and $D = 1 \times 10^{-5}\,\text{cm}^2\,\text{s}^{-1}$, the real net peak current depends linearly on the square-root of the frequency, over the frequency interval from 10 to 1000 Hz, whereas for $L < 2\,\mu\text{m}$ the dependence deviates from linearity. The peak current ratio of the forward and backward components is sensitive to the frequency. For instance, it varies from 1.19 to 1.45 over the frequency interval $10 \le f/\text{Hz} \le 1000$, which is valid for $L < 10\,\mu\text{m}$ and $D = 1 \times 10^{-5}\,\text{cm}^2\,\text{s}^{-1}$. It is important to emphasize that the frequency has no influence upon the peak potential of all components of the response and their values are virtually identical with the formal potential of the redox system.

For a quasireversible electrode reaction in thin-film voltammetry, the following numerical solution was obtained [156]:

$$\Psi_m = \frac{\kappa \exp(\alpha_a \varphi_m)\left[\frac{1}{\sqrt{f}} + [1 + \exp(-\varphi_m)]\sum_{j=1}^{m-1} \Psi_j S_{m-j+1}\right]}{\frac{1}{\sqrt{f}} - \kappa \exp(\alpha_a \varphi_m)S_1[1 + \exp(-\varphi_m)]} \tag{2.238}$$

where $\kappa = \frac{k_s}{\sqrt{Df}}$ is the dimensionless electrode kinetic parameter.

A quasireversible electrode reaction is controlled by the film thickness parameter Λ, and additionally by the electrode kinetic parameter κ. The definition and physical meaning of the latter parameter is the same as for quasireversible reaction under semi-infinite diffusion conditions (Sect. 2.1.2). Like for a reversible reaction, the dimensionless net peak current depends sigmoidally on the logarithm of the thickness parameter. The typical region of restricted diffusion depends slightly on κ. For instance, for $\log(\kappa) = -0.6$, the reaction is under restricted diffusion condition within the interval $\log(\Lambda) < 0.2$, whereas for $\log(\kappa) = 0.6$, the corresponding interval is $\log(\Lambda) < 0.4$.

Due to the similarity with surface electrode processes, a quasireversible reaction in thin-film voltammetry exhibits a quasireversible maximum and splitting of the net peak. The reasons causing these voltammetric features are the same as for surface

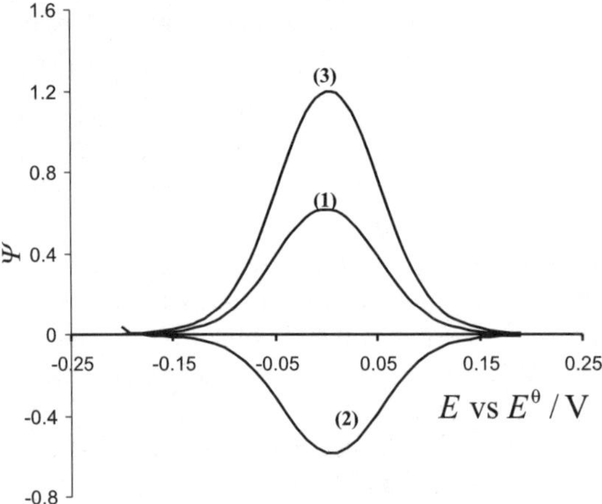

Fig. 2.93 Theoretical response of a reversible electrode reaction. The conditions of the simulations are: $\Lambda = 0.632$, $nE_{sw} = 50$ mV, $\Delta E = 10$ mV. The numbers 1, 2, and 3 designate the forward, backward and net component of the response

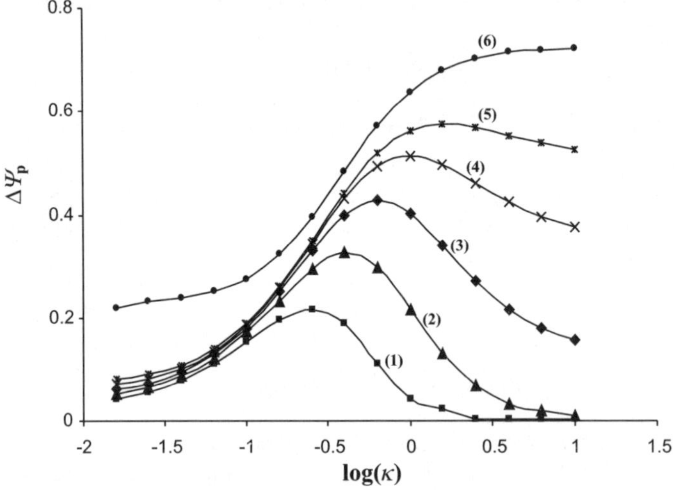

Fig. 2.94 Dependence of $\Delta\Psi_p$ on $\log(\kappa)$ for a film thickness parameter $\log(\Lambda) = -0.50$ (1); -0.33 (2); -0.20 (3); -0.10 (4) and -0.02 (5). *Curve* (6) corresponds to the quasireversible electrode reaction under semi-infinite diffusion conditions (Sect. 2.1.2). The other conditions of the simulations are: $nE_{sw} = 50$ mV, $\Delta E = 10$ mV, $\alpha_a = 0.5$ (with permission from [156])

processes (Sect. 2.5.1), i.e., they originate from the current sampling procedure in SWV and the rapid establishing of the equilibrium between redox species confined to the thin film. However, in the present case the quasireversible maximum and the splitting are also sensitive to the film thickness, which is a unique characteristic of an electrode reaction in a thin film. Figure 2.94 depicts the evolution of the quasi-reversible maxima for different values of the film thickness parameter. Increasing the thickness of the films shifts the maximum towards higher critical values of the electrode kinetic parameter. In parallel, the sharpness of the maximum decreases with growing film thickness. If the system is under semi-infinite diffusion conditions, the maximum vanishes, and the dependence $\Delta\Psi_p$ vs. $\log(\kappa)$ is represented by a sigmoidal curve (curve 6 in Fig. 2.94).

When a single electrode reaction is examined, the quasireversible maximum can be constructed by varying the frequency of the potential modulation, whereas the other parameters such as k_s, D, and L are typical constants for a given system. The frequency has a complex influence on the system, as it affects simultaneously the electrode kinetic parameter and the film thickness parameter. Figure 2.95 illustrates the effect of the frequency on the dimensionless net peak current for various standard rate constants (A) and film thicknesses (B). These results clearly show that the quasireversible maximum can be constructed by adjusting the frequency, and its position depends on the standard rate constant. The interrelation between the critical frequency (f_{max}) and the thickness of the film (L) is given by: $\log(f_{max}) = -0.99\log(L) + 1.73$ ($R = 0.99$). The later equation was found by analyzing the position of the quasireversible maximum for film thickness $1 \leq L/\mu m \leq 5$, $D = 1 \times 10^{-5}\,\mathrm{cm^2\,s^{-1}}$, $\log(k_s/\mathrm{cm\,s^{-1}}) = -2.6$, $\alpha_a = 0.5$, $nE_{sw} = 50\,\mathrm{mV}$, and $\Delta E = 10\,\mathrm{mV}$. Thus, this feature can be exploited for studying the electrode kinetics in a procedure analogous to that used for surface electrode processes (Chap. 2.5.1). Table 2.6 lists the linear regression lines for the dependence between the standard rate constant (k_s) and the critical frequency (f_{max}) of the quasireversible maximum for different film thicknesses.

Figure 2.96 shows the splitting of the net peak under increasing of the dimensionless electrode kinetic parameter for a given film thickness. The potential separation between split peaks increases in proportion to the electrode kinetic parameter and the amplitude of the potential modulation. The dependence of the peak potential separation on the amplitude is separately illustrated in Fig. 2.97. The analysis of the splitting by varying the amplitude is particularly appealing, since this instrumental parameter affects solely the split peak without altering the film thickness parameter. Table 2.7 lists the critical intervals of the film thickness and the electrode kinetic parameters attributed with the splitting.

Table 2.6 Linear regression functions representing the dependence of $\log(k_s)$ vs. $\log(f_{max})$ for different thickness of the film. The conditions of the simulations are: $nE_{sw} = 50\,\mathrm{mV}$, $\Delta E = 10\,\mathrm{mV}$, $\alpha_a = 0.5$, $D = 1 \times 10^{-5}\,\mathrm{cm^2\,s^{-1}}$. The last column lists the interval of k_s values over which the regression line was calculated

$L/\mu m$	Linear Regression Equation $\log(k_s)$ vs. $\log(f_{max})$	Interval of k_s values
1	$y = 1.05x - 4.43$	$-3.2 \leq \log(k_s) \leq -2.2$
2	$y = 0.97x - 4.00$	$-3 \leq \log(k_s) \leq -2.2$
3	$y = 1.02x - 3.87$	$-2.8 \leq \log(k_s) \leq -2.2$
4	$y = 1.42x - 4.20$	$-2.7 \leq \log(k_s) \leq -2.2$
5	$y = 1.67x - 4.33$	$-2.6 \leq \log(k_s) \leq -2.2$

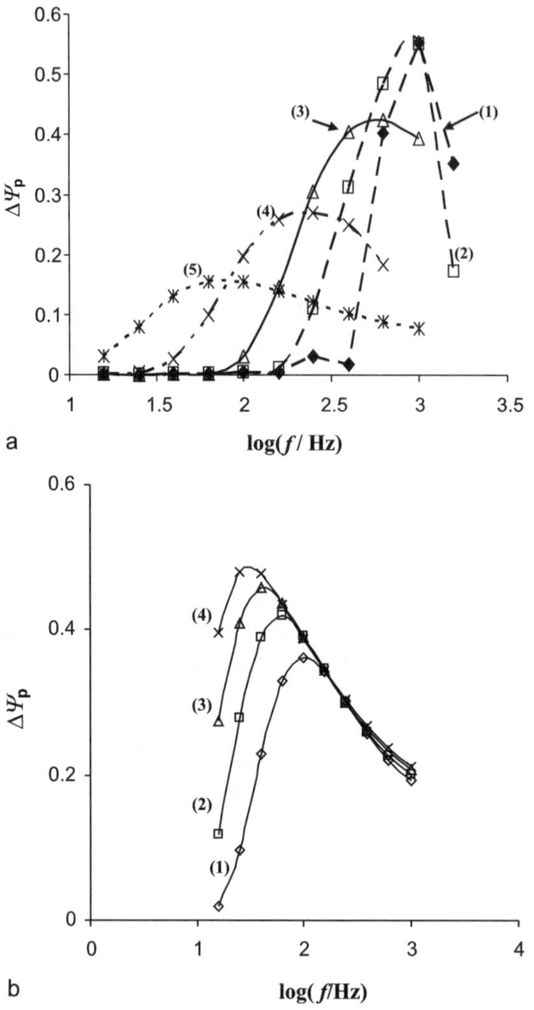

Fig. 2.95a,b Quasireversible maxima calculated by altering the frequency of the potential modulation for different standard rate constants (**a**) and thickness of the film (**b**). For (**a**), $L = 1\ \mu$m and $\log(k_s/\mathrm{cm\,s^{-1}}) = -1.5$ (*1*); -2 (*2*); -2.5 (*3*); -3 (*4*) and -3.5 (*5*). For (**b**), $\log(k_s/\mathrm{cm\,s^{-1}}) = -3$ and $L = 2$ (*1*); 3 (*2*); 4 (*3*) and 5 μm (*4*). The other conditions for both (**a**) and (**b**) are: $\alpha_a = 0.5$, $nE_{sw} = 50$ mV, $\Delta E = 10$ mV, $D = 1 \times 10^{-5}$ cm^2 s^{-1} (with permission from [156])

Similar to the surface electrode processes (Chap. 2.5.1) the peak current ratio of the split peaks ($\Psi_{p,c}/\Psi_{p,a}$) is a function of the electron transfer coefficient α_a. Note that the anodic and the cathodic peak is located at the more negative and more positive potentials, respectively. This type of dependence is given in Fig. 2.98. Over the interval $0.3 \leq \alpha_a \leq 0.7$ the dependence $\frac{\Psi_{p,c}}{\Psi_{p,a}}$ vs. α_a is linear, associated with the following linear regression line: $\frac{\Psi_{p,c}}{\Psi_{p,a}} = 3.0919\alpha_a - 0.4626$ ($R = 0.985$), which can be used for estimation of the electron transfer coefficient.

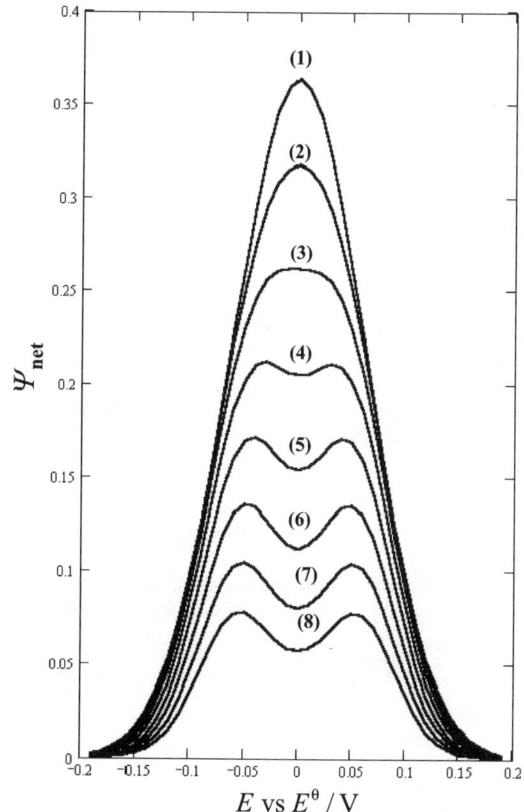

Fig. 2.96 Splitting of the net response under the influence of the electrode kinetic parameter. The electrode kinetic parameter is: $\log(\kappa) = -0.4$ (1); -0.3 (2), -0.2 (3); -0.1 (4); 0 (5); 0.1 (6); 0.2 (7) and 0.3 (8). The other conditions are: $\log(\Lambda) = -0.32$, $\alpha_a = 0.5$, $nE_{sw} = 80$ mV, $\Delta E = 10$ mV (with permission from [156])

Table 2.7 Typical intervals of κ associated with the splitting of the net peak, for different thickness of the thin film. The conditions of the simulations are: $\Delta E = 10$ mV, $\alpha_a = 0.5$

$nE_{sw} = 80$ mV	
$\log(\Lambda)$	$\log(\kappa)$
-0.597	> -0.5
-0.547	> -0.5
-0.500	> -0.4
-0.801	> -0.2
-0.324	> -0.1
-0.199	> 0.2
-0.102	No splitting

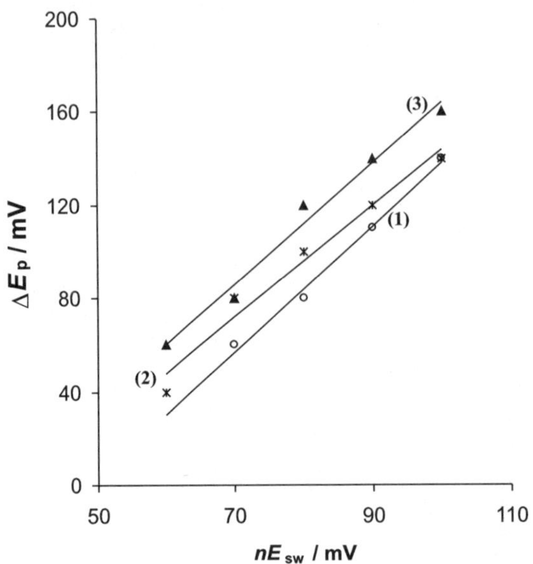

Fig. 2.97 Dependence of the potential separation of the split peaks on the amplitude for $\log(\kappa) = 0$ (1); 0.2 (2) and 0.4 (3). The other conditions are: $\log(\Lambda) = -0.324$, $\alpha_a = 0.5$, $\Delta E = 10$ mV (with permission from [156])

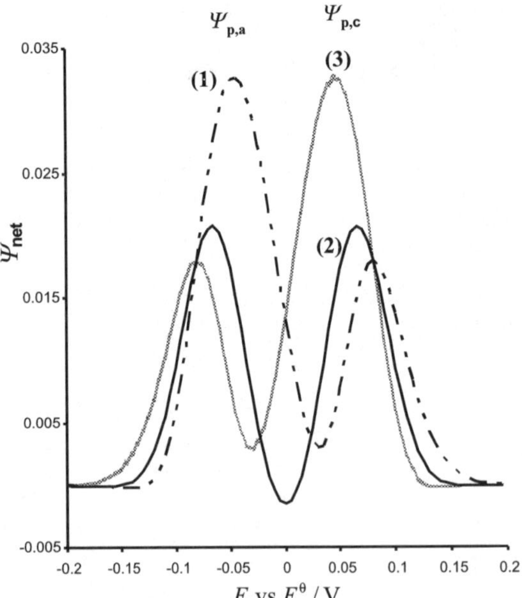

Fig. 2.98 Effect of the electron transfer coefficient on the split net peaks. The anodic electron transfer coefficient is: $\alpha_a = 0.3$ (1); 0.5 (2) and 0.7 (3). The other conditions are: $\log(\kappa) = -0.2$, $\log(\Lambda) = -0.398$, $nE_{sw} = 80$ mV, $\Delta E = 5$ mV (with permission from [156])

References

1. Osteryoung JG, Osteryoung RA (1985) Anal Chem 57:101 A
2. Aoki K, Maeda K, Osteryoung J (1989) J Electroanal Chem 272:17
3. Zachowski EJ, Wojciechowski M, Osteryoung J (1986) Anal Chim Acta 183:47
4. Krulic D, Fatouros N, El Belamachi MM (1995) J Electroanal Chem 385:33
5. Brookes BA, Ball JC, Compton RG (1999) J Phys Chem B 103:5289
6. Molina A, Serna C, Camacho L (1995) J Electroanal Chem 394:1
7. Osteryoung J, O'Dea JJ (1986) Square-wave voltammetry. In: Bard AJ (ed) Electroanalytical chemistry, vol 14. Marcel Dekker, New York, p 209
8. Scholz F, Nitschke L, Henrion G (1988) Fresenius Z Anal Chem 332:805
9. Pelzer J, Scholz F, Henrion G, Nitschke L (1989) Electroanalysis 1:437
10. Scholz F, Draheim M, Henrion G (1990) Fresenius Z Anal Chem 336:136
11. Ramaley L, Krause MS Jr (1969) Anal Chem 41:1362
12. Lovrić M (1994) Ann Chim 84:379
13. Pižeta I, Zelić M (1996) Anal Chim Acta 319:31
14. Mirčeski V, Lovrić M (2001) J Electroanal Chem 497:114
15. O'Dea JJ, Osteryoung J, Osteryoung RA (1981) Anal Chem 53:695
16. Fatouros N, Krulic D (1998) J Electroanal Chem 443:262
17. Brookes BA, Compton RG (1999) J Phys Chem B 103:9020
18. Nuwer MJ, O'Dea JJ, Osteryoung J (1991) Anal Chim Acta 251:13
19. O'Dea JJ, Osteryoung J, Osteryoung RA (1983) J Phys Chem 87:3911
20. O'Dea JJ, Osteryoung J, Lane T (1986) J Phys Chem 90:2761
21. Go WS, O'Dea JJ, Osteryoung J (1988) J Electroanal Chem 255:21
22. Fatouros N, Krulic D (2002) J Electroanal Chem 520:1
23. Zelić M (2006) Croat Chem Acta 79:49
24. Fatouros N, Krulic D, Larabi N (2003) J Electroanal Chem 549:81
25. Fatouros N, Krulic D, Larabi N (2004) J Electroanal Chem 568:55
26. Komorsky-Lovrić Š, Lovrić M, Bond AM (1993) Electroanalysis 5:29
27. Brookes BA, Macfie G, Compton RG (2000) J Phys Chem B 104:5784
28. Aoki K, Tokuda K, Matsuda H, Osteryoung J (1986) J Electroanal Chem 207:25
29. O'Dea JJ, Wojciechowski M, Osteryoung J, Aoki K (1985) Anal Chem 57:954
30. Ramaley L, Tan WT (1987) Can J Chem 65:1025
31. Whelan DP, O'Dea JJ, Osteryoung J, Aoki K (1986) J Electroanal Chem 202:23
32. Bond AM, Henderson TLE, Mann DR, Mann TF, Thormann W, Zoski CG (1988) Anal Chem 60:1878
33. Kadish KM, Ding JQ, Malinski T (1984) Anal Chem 56:1741
34. Heusler KE (1982) In: Bard AJ (Ed) Encyclopedia of electrochemistry of the elements, vol IX, Part A. Marcel Dekker, New York, p 229
35. Kounaves SP, Deng W (1993) Anal Chem 65:375

36. Fatouros N, Krulic D, Lopez-Tenes M, El Belamachi MM (1996) J Electroanal Chem 405:197
37. Singleton ST, O'Dea JJ, Osteryoung J (1989) Anal Chem 61:1211
38. Murphy MM, O'Dea JJ, Osteryoung J (1991) Anal Chem 63:2743
39. Tallman DE (1994) Anal Chem 66:557
40. Kounaves SP, O'Dea JJ, Chandrasekhar P, Osteryoung J (1986) Anal Chem 58:3199
41. Mlakar M, Lovrić M (1990) Analyst 115:45
42. Lovrić M (2002) Kem Ind (Zagreb) 51:205
43. Miles AB, Compton RG (2000) J Electroanal Chem 487:75
44. Penczek M, Stojek Z (1986) J Electroanal Chem 213:177
45. Kounaves SP, O'Dea JJ, Chandrasekhar P, Osteryoung J (1987) Anal Chem 59:386
46. Ball JC, Compton RG (1998) J Phys Chem B 102:3967
47. Wikiel K, Osteryoung J (1989) Anal Chem 61:2086
48. Wechter C, Osteryoung J (1989) Anal Chem 61:2092
49. Kounaves SP, Deng W (1991) J Electroanal Chem 306:111
50. Agra-Gutierrez C, Ball JC, Compton RG (1998) J Phys Chem B 102:7028
51. Powell M, Ball JC, Tsai Y-C, Suarez MF, Compton RG (2000) J Phys Chem B 104:8268
52. Nicholson RS, Olmstead ML (1972) Numerical solutions of integral equations. In: Matson JS, Mark HB, MacDonald HC (eds) Electrochemistry: calculations, simulations and instrumentation, vol 2. Marcel Dekker, New York, p 119
53. Zeng J, Osteryoung RA (1986) Anal Chem 58:2766
54. O'Dea JJ, Wikiel K, Osteryoung J (1990) J Phys Chem 94:3628
55. Osteryoung J, O'Dea JJ (1986) Square-wave voltammetry. In: Bard AJ (ed) Electroanalytical chemistry, vol 14. Marcel Dekker, New York, p 209
56. Miles AB, Compton RG (2000) J Phys Chem B 104:5331
57. Miles AB, Compton RG (2001) J Electroanal Chem 499:1
58. Rudolph M (2001) J Electroanal Chem 503:15
59. Molina A (1998) J Electroanal Chem 443:163
60. Garay F, Lovrić M (2002) J Electroanal Chem 518:91
61. Krulic D, Larabi N, Fatouros N (2005) J Electroanal Chem 579:243
62. Garay F, Lovrić M (2002) Electroanalysis 14:1635
63. Garay F, Lovrić M (2002) J Electroanal Chem 527:85
64. Molina A, Serna C, Martínez-Ortiz F (2000) J Electroanal Chem 486:9
65. Correia dos Santos MM, Simões Gonçalves ML, Romão JC (1996) J Electroanal Chem 413:97
66. Geary CD, Weber SG (2003) Anal Chem 75:6560
67. Geary CD, Zudans I, Goponenko AV, Asher SA, Weber SG (2005) Anal Chem 77:185
68. Macfie G, Atherton JH, Compton RG (2002) Electroanalysis 14:479
69. Macfie G, Brookes BA, Compton RG (2001) J Phys Chem B 105:12534
70. Bragato C, Daniele S, Antonietta Baldo M (2005) Electroanalysis 17:1370
71. Abreu FC, Goulart MOF, Oliveira Brett AM (2002) Electroanalysis 14:29
72. Mirčeski V, Gulaboski R, Scholz F (2002) Electrochem Commun 4:814
73. Smith DE (1963) Anal Chem 35:602
74. Nicholson RS, Shain I (1965) Anal Chem 37:190
75. Kim M-H, Birke RL (1983) Anal Chem 55:522
76. Reeves JH, Song S, Bowden EF (1993) Anal Chem 65:683
77. O'Dea JJ, Osteryoung JG (1993) Anal Chem 65:3090
78. Saccucci TM, Rusling JF (2001) J Phys Chem B 105:6142
79. Komorsky-Lovrić Š, Lovrić M (1995) J Electroanal Chem 384:115
80. Lovrić M (1991) Elektrokhimija 27:186
81. Komorsky-Lovrić Š, Lovrić M (1995) Kem Ind 44:355
82. Komorsky-Lovrić Š, Lovrić M (1995) Anal Chim Acta 305:248
83. Zhang J, Guo Si-X, Bond AM, Honeychurch MJ, Oldham KB (2005) J Phys Chem B 109:8935
84. Mirčeski V, Lovrić M (1997) Electroanalysis 9:1283

85. Mirčeski V, Lovrić M, Gulaboski R (2001) J Electroanal Chem 515:91
86. Mirčeski V, Lovrić M, (2000) Croat Chem Acta 73:305
87. Mirčeski V, Gulaboski R (2001) Electroanalysis 13:1326
88. Mirčeski V, Gulaboski R (2003) J Solid State Electrochem 7:157
89. Gulaboski R, Mirčeski V, Lovrić M, Bogeski I (2005) Electrochem Commun 7:515
90. O'Dea JJ, Osteryoung JG (1997) Anal Chem 69:650
91. Mirčeski V, Gulaboski R (2003) Croat Chem Acta 76:37
92. Lovrić M, Komorsky-Lovrić Š, (1988) J Electroanal Chem 248:239
93. Komorsky-Lovrić Š (1996) Fresenius J Anal Chem 356:306
94. Mirčeski V, Lovrić M, Jordanoski B (1999) Electroanalysis 11:660
95. Quentel F, Elleouet C (2001) Electroanalysis 13:1030
96. Quentel F, Mirčeski V, Laouenan A, Elleouet C, Madec CL (2003) Electroanalysis 15:270
97. Komorsky-Lovrić Š (2000) J Electroanal Chem 482:222
98. Mirčeski V, Gulaboski R, Petrovska-Jovanović S, Stojanova K (2001) Portugaliae Electrochim Acta 19:25
99. Liu A, Wei M, Honma I, Zhou H (2005) Anal Chem 77:8068
100. Jeuken LJC, Jones AK, Chapman SK, Cecchini G, Armstrong FA (2002) J Am Chem Soc 124:5702
101. Jeuken LJC, McEvoy JP, Armstrong FA (2002) J Phys Chem B 106:2304
102. Marchiando NC, Zón MA, Fernández H (2003) Electroanalysis 15:40
103. Molina PG, Zón MA, Fernández H (2000) Electroanalysis 12:791
104. Mirčeski V, Gulaboski R (2002) Mikrochim Acta 138:33
105. Quentel F, Mirčeski V (2003) Electroanalysis 15:1787
106. Laviron E, Roullier L (1980) J Electroanal Chem 115:65
107. Mirčeski V, Jordanoski B, Avramov-Ivić M (1998) J Serb Chem Soc 63:719
108. Quentel F, Mirčeski V (2004) Electroanalysis 16:1690
109. Mugweru A, Rusling JF (2001) Electrochem Commun 3:406
110. Lovrić M, Branica M (1987) J Electroanal Chem 226:239
111. Lovrić M, Komorsky-Lovrić Š, Murray RW (1988) Electrochim Acta 33:739
112. Komorsky-Lovrić Š, Lovrić M (1989) Fresenius Z Anal Chem 335:289
113. Lovrić M, Komorsky-Lovrić Š, Bond AM (1991) J Electroanal Chem 319:1
114. Komorsky-Lovrić Š, Lovrić M, Branica M (1992) J Electroanal Chem 335:297
115. O'Dea JJ, Ribes A, Osteryoung JG (1993) J Electroanal Chem 345:287
116. Garay F (2003) J Electroanal Chem 548: 1
117. Garay F, Solis V, Lovrić M (1999) J Electroanal Chem 478:17
118. Garay F (2001) J Electroanal Chem 505:100
119. Garay F, Solis VM (2004) Electroanalysis 16:450
120. Lovrić M (2002) Electroanalysis 14:405
121. Komorsky-Lovrić Š (1987) J Serb Chem Soc 52:43
122. Garay F, Solis VM (2003) J Electroanal Chem 544:1
123. Garay F, Solis VM (2001) J Electroanal Chem 505:109
124. Jelen F, Tomschik M, Paleček E (1997) J Electroanal Chem 423:141
125. Çakir S, Biçer E, Çakir O (2000) Electrochem Commun 2:124
126. Roque da Silva AMS, Lima JC, Oliva Teles MT, Oliveira Brett AM (1999) Talanta 49:611
127. Komorsky-Lovrić Š, Branica M (1994) Fresenius J Anal Chem 349:633
128. Mirčeski V (2001) J Electroanal Chem 508:138
129. Mirčeski V, Lovrić M (2004) J Electroanal Chem 565:191
130. Mirčeski V, Quentel F (2005) J Electroanal Chem 578:25
131. Brainina KhZ (1971) Talanta 18:513
132. Florence TM (1979) J Electroanal Chem 97:219
133. Lovrić M, Pižeta I, Komorsky-Lovrić Š (1992) Electroanalysis 4:327
134. Mirčeski V, Lovrić M (1998) Electroanalysis 10:976
135. Mirčeski V, Lovrić M (1999) Electroanalysis 11:984
136. Mirčeski V, Lovrić M (1999) Anal Chim Acta 386:47
137. Gulaboski R, Mirčeski V, Komorsky-Lovrić Š (2002) Electroanalysis 14:345

138. Gulaboski R, Mirčeski V, Komorsky-Lovrić Š, Lovrić M (2004) Electroanalysis 16:832
139. Schultze JW, Koppitz FD (1976) Electrochim Acta 21:327
140. Harman AR, Baranski AS (1991) Anal Chem 63:1158
141. Luther GW, Swartz CB, Ullman WJ (1988) Anal Chem 60:1721
142. Wong GTF, Zhang LS (1992) Talanta 39:355
143. Hernández P, Ballesteros Y, Galán F, Hernández L (1994) Electroanalysis 6:51
144. Zhang JZ, Millero FJ (1994) Anal Chim Acta 284:497
145. Li H, Smart RB (1996) Anal Chim Acta 325:25
146. Barra CM, Correia dos Santos MM (2001) Electroanalysis 13:1098
147. Çakir S, Biçer E, Çakir O (1999) Electrochem Commun 1:257
148. Lovrić M, Mlakar M (1995) Electroanalysis 7:1121
149. Mladenov M, Mirčeski V, Gjorgoski I, Jordanoski B (2004) Bioelectrochem 65:69
150. Garay F (2003) J Electroanal Chem 548:11
151. Mirčeski V, Gulaboski R, Jordanoski B, Komorsky-Lovrić Š (2000) J Electroanal Chem 490:37
152. Temeterk YM, Kamal MM, Ahmed GA-W, Ibrahim HSM (2003) Anal Sci 19:1115
153. Sladkov V, Bolyos A, David F (2002) Electroanalysis 14:128
154. Aoki K, Osteryoung J (1988) J Electroanal Chem 240:45
155. Mirčeski V, Gulaboski R, Scholz F (2004) J Electroanal Chem 566:351
156. Mirčeski V (2004) J Phys Chem B 108:13719
157. Majid E, Hrapovic S, Liu Y, Male KB, Luong JHT (2006) Anal Chem 78:762
158. Kumar V, Heineman WR (1987) Anal Chem 59:842
159. Alcantara K, Munge B, Pendon Z, Frank HA, Rusling JF (2006) J Am Chem Soc 128:14930
160. Quentel F, Mirčeski V, L'Her M, (2005) Anal Chem 77:1940
161. Mirčeski V, Quentel F, L'Her M, Pondaven A (2005) Electrochem Commun 7:1122
162. Mirčeski V (2003) J Electroanal Chem 545:29

Chapter 3
Applications

3.1 Quantitative Analysis

Electroanalytical application of square-wave voltammetry can be divided into direct and stripping measurements. Some ions and compounds that were analyzed directly, i.e., without accumulation of the reactant or product of the electrode reaction, are listed in Table 3.1. The stripping methods are based either on the accumulation of amalgams and metal deposits, or on the adsorptive accumulation of organic substances and metal complexes. For the trace determination of zinc, cadmium, lead and copper ions, anodic stripping SWV on various types of thin mercury film covered electrodes was applied [35–51]. Mercury film covered carbon fiber [52–56] and iridium [57] microelectrodes were also used. Figure 3.1 displays square-wave stripping voltammograms at a mercury-coated nanoband electrode for solutions of increasing cadmium and lead concentrations following a 2 min deposition [46]. The inset shows the linear relationships between peak currents and the metal concentrations. The band electrode was fabricated from ultrathin carbonized polyacrylonitrile films. The volume of electrolyte in the cell was 1 µl. The limit of detection of lead ions was found to be 1×10^{-9} M.

The following elements were measured by using metal deposits on solid electrodes: bismuth on a glassy carbon electrode (GCE) [58], cadmium on a boron-doped diamond electrode [59], a bismuth-film-covered pencil-lead electrode [60] and a carbon paste electrode modified with bismuth powder [61], copper on the GCE [62–64], gold on chemically modified electrodes [65], indium on bismuth-film-covered electrodes [66], lead on a carbon fiber microelectrode [67], a boron-doped diamond electrode [68], the GCE [69, 70] and a gold electrode [71], mercury on a screen-printed electrode [72], chemically modified GCE [73] and gold electrode [74] and a carbon paste electrode modified with silica [75], nickel on the GCE [76], thallium on a graphite microelectrode [77] and zinc on the GCE [78]. Figure 3.2 shows square-wave stripping voltammograms ($f = 25$ Hz, $E_{sw} = 15$ mV and $\Delta E = 1.5$ mV) of copper on glassy carbon electrode following an insonated deposition of 60 s at -1.5 V vs. SCE [64]. During the deposition step ultrasound

V. Mirčeski, Š. Komorsky-Lovrić, M. Lovrić, *Square-Wave Voltammetry*
doi: 10.1007/978-3-540-73740-7, ©Springer 2008

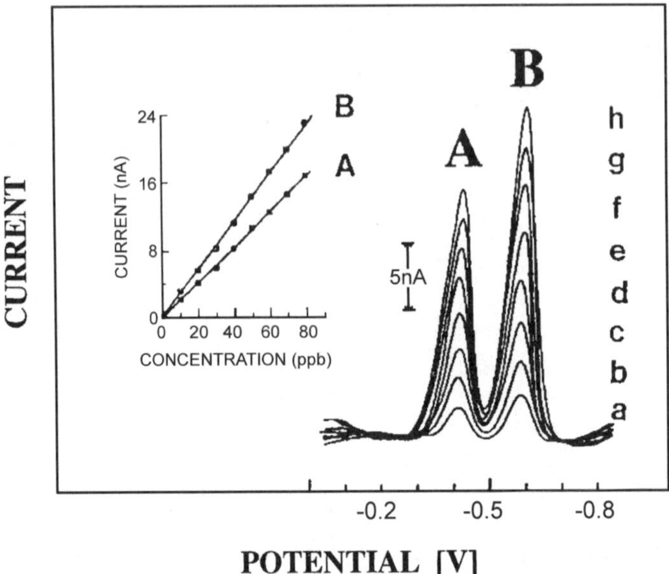

Fig. 3.1 Stripping voltammograms for mixtures containing increasing levels of cadmium and lead, 10–80 µg/l ($a - h$) in 0.1 M acetate buffer (pH 4.2). Deposition for 2 min at -1.0 V vs. Ag/AgCl/KCl(sat.), using unstirred solutions; $f = 45$ Hz, $E_{sw} = 20$ mV and $\Delta E = 6$ mV (reprinted from [46] with permission)

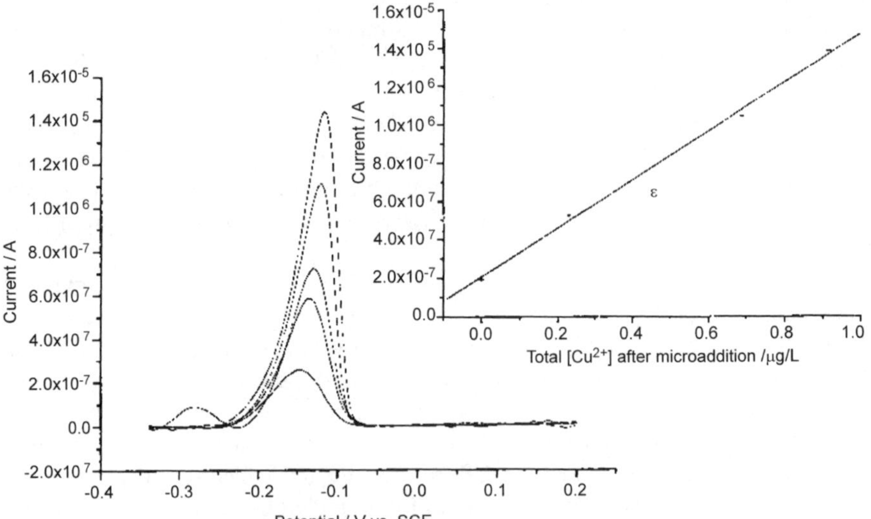

Fig. 3.2 Sono-square-wave anodic stripping voltammetric traces for an insonated deposition of 60 s at -1.5 V. *Traces* show background corrected standard additions to sono-solvent extracted laked horse blood solution (test solution 0.05%, by volume–blood). Each 10 µl addition corresponds to an increase in copper concentration of 0.22 µg/l. *Calibration graph* shown inset ($R^2 = 0.9972$) gives concentration of 1.637 mg/l (reprinted from [64] with permission)

Table 3.1

Compound	Limit of detection/M	The linear range/M	Reference
Ametryne	2×10^{-7}	5×10^{-7}–1×10^{-5}	[1]
5-aminosalicylic acid	5×10^{-7}	1×10^{-6}–6×10^{-5}	[2]
Amisulpride	1×10^{-6}	4×10^{-6}–6×10^{-4}	[3]
Arsenic	1×10^{-8}	5×10^{-8}–1×10^{-6}	[4]
Atrazine	2×10^{-8}	1×10^{-7}–2.5×10^{-6}	[5]
Azidothymidine	1×10^{-9}	1×10^{-8}–5×10^{-7}	[6]
Benzoylecgonine	5×10^{-7}	1×10^{-6}–4×10^{-5}	[7, 8]
Bipyridinium oximes	2×10^{-8}	2×10^{-8}–5×10^{-6}	[9]
Caffeine	2×10^{-6}	1×10^{-5}–2.5×10^{-4}	[10]
Cefixime	3×10^{-6}	6×10^{-6}–2×10^{-4}	[11]
Chlorine	7×10^{-6}	1.7×10^{-5}–5×10^{-4}	[12]
Dopamine	5×10^{-7}	1×10^{-6}–5×10^{-5}	[13]
Ethanol	3×10^{-3}	2×10^{-2}–1.7	[14]
Fenofibrate	1×10^{-7}	4×10^{-7}–1.4×10^{-5}	[15]
Formoterol	5×10^{-6}	8×10^{-6}–6×10^{-5}	[16]
Indium(III)	8×10^{-8}	1.6×10^{-6}–3×10^{-4}	[17]
Lead(II)	5×10^{-7}	2×10^{-6}–1×10^{-4}	[18, 19]
Manganese(II)	3×10^{-7}	5×10^{-7}–1×10^{-5}	[20]
Niobium(III)	2×10^{-6}	5×10^{-6}–3×10^{-4}	[21]
Nitroaromatic and nitramine explosives	2×10^{-5}	1×10^{-3}–3×10^{-3}	[22]
Riboflavin	5×10^{-8}	1×10^{-7}–1×10^{-5}	[23]
Serotonin	2×10^{-9}	1×10^{-8}–5×10^{-7}	[24]
L-tyrosine	4×10^{-7}	2×10^{-6}–5×10^{-4}	[25]
Uranium(VI)	7×10^{-7}	1×10^{-6}–1×10^{-4}	[26, 27]
Uric acid	5×10^{-5}	1×10^{-4}–1×10^{-3}	[28]
Valacyclovir	5×10^{-8}	4×10^{-6}–2×10^{-4}	[29]
Vanillin	4×10^{-7}	5×10^{-6}–4×10^{-4}	[30, 31]
Vardenafil	1×10^{-7}	4×10^{-7}–2×10^{-5}	[32]
Venlafaxine	5×10^{-7}	1×10^{-6}–5×10^{-6}	[33]
Vitamins b_6 and b_{12}	5×10^{-5}	1×10^{-4}–1×10^{-3}	[34]

was effective in maintaining a clean electrode and promoting the mass transport of copper ions. The method was combined with ultrasonically enhanced solvent extraction of copper from a horse blood. Figure 3.3 shows the determination of traces of mercury ions using square-wave anodic stripping voltammetry on glassy carbon electrode coated with Kryptofix-222 [73]. The modification of GCE was accomplished by spin-coating with a methanolic solution of Kryptofix-222, followed by the rinsing with water. The deposition potential and time were $-0.5\,\mathrm{V}$ vs. Ag/AgCl and 300 seconds, respectively, the SW frequency was $15\,\mathrm{Hz}$ and SW amplitude was $50\,\mathrm{mV}$. A detection limit as low as $1 \times 10^{-12}\,\mathrm{M}$ was obtained.

For the analysis of surface-active, electroactive organic compounds, the adsorptive stripping SWV was used. The method was applied to numerous analytes. Several of them are listed in Table 3.2. Some examples of metal complexes which were used for the quantitative analysis of metal ions by adsorptive

Table 3.2

Compound	Limit of detection / M	The linear range / M	Reference
Abscisic acid	2×10^{-10}	3×10^{-10}–2×10^{-9}	[79]
Adriamycin	8×10^{-11}	1×10^{-10}–1×10^{-9}	[80]
Amlodipine besylate	1.4×10^{-8}	4×10^{-8}–2×10^{-6}	[81]
Azithromycin	6×10^{-10}	6×10^{-10}–9×10^{-9}	[82]
Azobenzene	3×10^{-12}	1×10^{-10}–1×10^{-7}	[83]
Azosalicylic acids	1×10^{-9}	5×10^{-9}–1×10^{-7}	[84]
1,4-benzodiazepines	1.2×10^{-8}	1×10^{-7}–1×10^{-5}	[85]
Berberine	5×10^{-8}	1×10^{-7}–2×10^{-6}	[86, 87]
Captopril	2×10^{-9}	2.5×10^{-9}–9×10^{-7}	[88, 89]
Cefazolin	2.6×10^{-10}	1×10^{-8}–5×10^{-7}	[90]
Cefonicid	4×10^{-8}	1×10^{-7}–1×10^{-6}	[91]
Cercosporin	1×10^{-7}	5×10^{-7}–5×10^{-6}	[92]
Cimetidine	4×10^{-9}	1×10^{-8}–8×10^{-6}	[93]
Cyclofenil	1.5×10^{-8}	5×10^{-8}–6×10^{-6}	[94]
Danazol	6×10^{-9}	7×10^{-8}–4×10^{-7}	[95]
Dimethoate	4×10^{-9}	4×10^{-9}–3×10^{-8}	[96]
Doxazosin	2×10^{-11}	5×10^{-11}–5×10^{-9}	[97]
Epinephrine	1×10^{-8}	1×10^{-7}–1×10^{-5}	[98, 99]
Ethinylestradiol	6×10^{-10}	2×10^{-9}–6×10^{-7}	[100]
Ethylenediaminetetraacetic acid	3×10^{-7}	6×10^{-7}–5×10^{-5}	[101]
Famotidine	1×10^{-7}	5×10^{-7}–5×10^{-6}	[102]
Fluoxetine	4×10^{-8}	5×10^{-7}–5×10^{-6}	[103]
Fluvoxamine	5×10^{-9}	6×10^{-9}–3×10^{-7}	[104]
Glucose	2×10^{-6}	2×10^{-5}–1.4×10^{-3}	[105]
1-hydroxypyrene	1×10^{-9}	5×10^{-9}–4×10^{-7}	[106]
Imidacloprid	1.6×10^{-8}	2×10^{-8}–5×10^{-7}	[107]
Lamotrigine	3×10^{-9}	5×10^{-9}–2×10^{-8}	[108]
Ketorolac	1×10^{-11}	1×10^{-10}–1×10^{-8}	[109]
Melatonin	3×10^{-10}	1×10^{-9}–5×10^{-8}	[110]
Meloxicam	5×10^{-10}	1×10^{-9}–1×10^{-7}	[111]
Metamitron	3.7×10^{-7}	8×10^{-7}–8×10^{-6}	[112]
Midazolam	2×10^{-8}	5×10^{-8}–1×10^{-6}	[113]
Mifepristone	5×10^{-9}	2×10^{-8}–6×10^{-7}	[114]
Moxifloxacin	4×10^{-8}	4×10^{-7}–1×10^{-5}	[115]
Nicotinamide adenine dinucleotide	7×10^{-9}	1×10^{-8}–5×10^{-6}	[116, 117]
Nifedipine	1.2×10^{-9}	3×10^{-9}–3.6×10^{-7}	[118]
4-nitrophenol	2×10^{-8}	3×10^{-6}–5×10^{-5}	[119]
Nitroxynil	8.4×10^{-10}	3×10^{-9}–2×10^{-7}	[120]
Norfloxacin	3×10^{-6}	1×10^{-5}–2×10^{-4}	[121]
Paraquat	1×10^{-7}	5×10^{-7}–1×10^{-5}	[122, 123]
Pefloxacin	4.5×10^{-10}	1×10^{-9}–1×10^{-7}	[124]
Pentachlorophenol	2×10^{-8}	1×10^{-7}–6×10^{-5}	[125]
Pantoprazole	5×10^{-10}	1×10^{-9}–5×10^{-8}	[126]
Sildenafil citrate	1×10^{-8}	4×10^{-8}–5×10^{-7}	[127]
Sulfamethazine	6.8×10^{-9}	1×10^{-8}–1×10^{-6}	[128]
Sulphamethoxypyridazine	2×10^{-8}	1×10^{-7}–2×10^{-6}	[129]
Tarabine PFS	1.1×10^{-10}	2×10^{-10}–6×10^{-8}	[130]
Tetramethrin	8.5×10^{-9}	3×10^{-8}–3×10^{-7}	[131]
Tianeptine	2×10^{-8}	8×10^{-8}–8×10^{-7}	[132]
Trimetazidine	2×10^{-8}	5×10^{-8}–5×10^{-6}	[133]
Warfarin	6.5×10^{-10}	5×10^{-9}–4×10^{-7}	[134]
Xanthine	5×10^{-10}	1×10^{-9}–6×10^{-8}	[135]
Xanthosine	5×10^{-8}	1×10^{-7}–8×10^{-7}	[135]

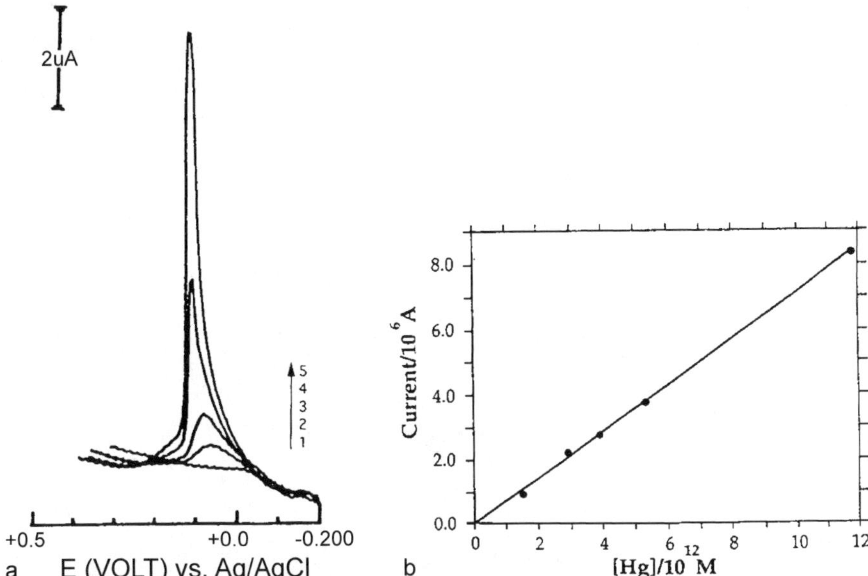

Fig. 3.3 a Anodic Osteryoung square wave stripping voltammograms of a GCE coated with Kryptofix-222 in a solution of 0.01 M acetate buffer (pH 4.0): (1) Hg(II) free; (2) 1.51×10^{-12} M Hg(II); (3) 2.89×10^{-12} M Hg(II); (4) 5.34×10^{-12} M Hg(II); (5) 1.18×10^{-11} M Hg(II). **b** Anodic stripping peak currents of mercury as a function of mercury(II) concentration. Potential and time of deposition equal -0.5 V and 300 s, respectively (reprinted from [73] with permission)

stripping SWV are the following: aluminium(III)-cupferron [136], beryllium(II)-arsenazo-I [137], bismuth(III)-morin [138], cadmium(II)-oxine [139], cadmium(II)-carbamoyl-phosphonic acid [140], chromium(III)-triethylenetetranitrilohexaacetic acid [141], cobalt(II)-dioxime [142], cobalt(II)-dimethylglyoxime [143], copper(II)-thiourea [144], copper(II)-5,5-dimethylcyclohexane-1,2,3-trione-1,2-dioxime-3-thio-semicarbazone (Cu-DCDT) [145], europium(III)-2-thenoyltrifluoroacetone [146], europium(III)-salicylate [147], molybdenum(VI)-2,3-dihydroxynaphthalene [148], nickel(II)-dimethylglyoxime (Ni-DMG) [149–151], platinum(II)-dimethyl-glyoxime [152], platinum(II)-formazone [153], titanium(IV)-methylthymol blue [154], uranium(VI)-humate [155], uranium(VI)-4-(2-hydroxyethyl)-1-piprazine-ethane-sulfonic acid [156] and vanadium(V)-chloranilic acid [157]. Figure 3.4 shows cathodic stripping square-wave voltammograms of copper(II)-DCDT complex adsorbed on the surface of hanging mercury drop electrode [145]. The method was used for the determination of copper in olive oil samples. The detection limit was 8×10^{-9} M. A cathodic stripping SWV of nickel(II)-DMG complex adsorbed on the surface of a thin mercury film covered glassy carbon rotating disk electrode are shown in Figure 3.5. The supporting electrolyte was NH_3/NH_4Cl buffer pH 9, and the concentration of DMG was 1×10^{-3} M. The adsorptive accumulation of the complex was performed at -0.7 V vs. Ag/AgCl, for 60 seconds, with the electrode rotation speed of 600 r.p.m. In the stripping step the frequency was 40 Hz, the

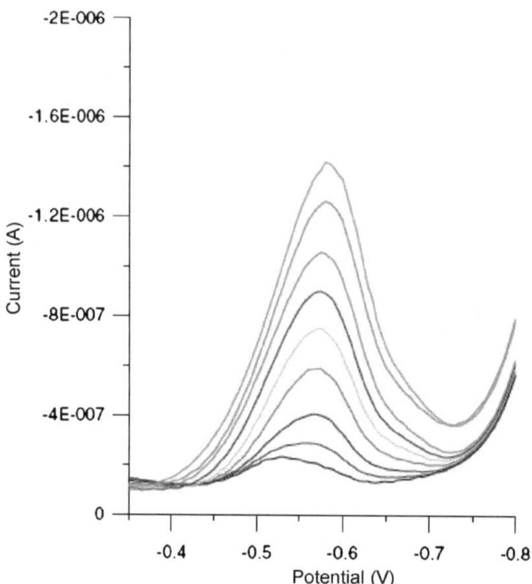

Fig. 3.4 Ad-SSWV voltammograms of Cu-DCDT obtained for Cu concentrations between 0 to 35 ng/ml ($8 \times 10^{-5}\%$ DCDT, 1.6% ethanol and 1 M HCl). $E_{acc} = -0.350$ V; $t_{acc} = 60$ s, frequency $= 200$ Hz; step potential $= 5$ mV; amplitude $= 60$ mV (reprinted from [145] with permission)

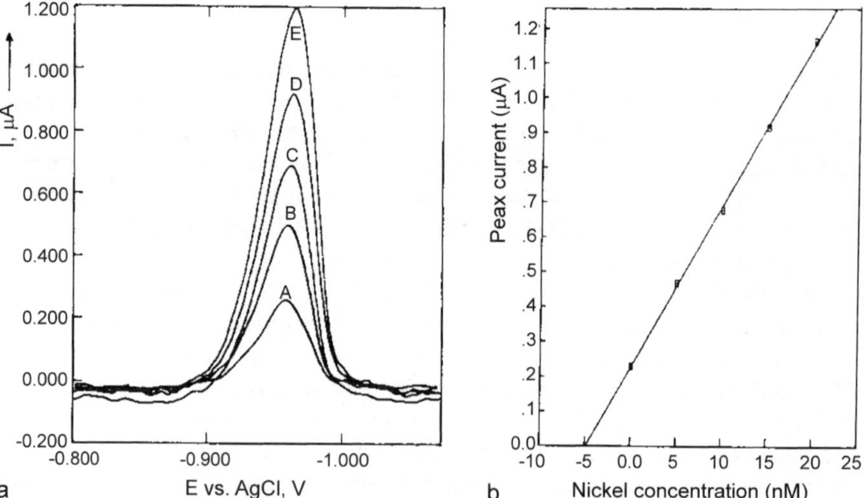

Fig. 3.5 a Background subtracted SW stripping voltammograms for Ni(II) adsorbed on a MFE as the Ni(II)-DMG complex: (A) 5 nmol/l Ni(II); (B) , (C), (D) and (E) standard additions of 5, 10, 15 and 20 nmol/l Ni(II), respectively. **b** Standard addition plot for the standard additions of **a** (reprinted from [149] with permission)

amplitude was 10 mV and the scan increment was 2 mV. The limit of detection was 8.5×10^{-10} M and the linear range was from 1×10^{-9} to 1×10^{-6} M.

Ion-exchange reactions were used for the accumulation of europium(III) [158] and iron(III) [159] ions on the surface of GCE coated with Nafion®, and chromium(VI) ions on the surface of GCE covered by a pyridine-functionalized sol-gel film [160], which were combined with the stripping SWV. Furthermore, a cathodic stripping SWV was used for the determination of sulfide [161, 162], thiols [163–166], selenium(IV) [167–170], halides [171–173] and arsenic [174] accumulated on the surface of mercury electrode.

3.2 Qualitative Identification of Phases

SWV was used for the qualitative identification of solid phases [175–188]. The method is based on mechanical immobilization of microparticles of insoluble organic or inorganic compounds on the surface of a suitable solid electrode, e.g. on a paraffin impregnated graphite rod which is then used as the working electrode in a SWV experiment [189]. Depending on the electrochemical properties of microparticles, the electrode reaction may lead to dissolution of the particle [175] or to its transformation into another solid phase. The latter may in some cases proceed by insertion of cations, or anions of aqueous electrolyte into the structure of microcrystals [176]. The method was used for the detection of manganese in carbonates [177] and marine sediments [178], cobalt, copper, antimony, tin, iron, zinc and zirconium in archaeological glass [179], boron and lead in minerals and ceramic materials [180, 181] and tin, copper, lead and nickel in bronzes [182]. Also, it was used for identification of various organic solids, such as quinhydrone, indigo, acridine, famotidine, probucol, thionicotinoylanilide [183], cocaine [184], benzocaine, cinchocaine, lidocaine, procaine, codeine [185], carmine, cochineal red, naphthoquinone dyes, anthraquinone dyes, flavone dyes [186], 5-aminosalicylic acid, ciprofloxacin, azithromycin [187] and simvastatin [188].

Square-wave voltammograms of microparticles of three quinone dyes (lawson, alizarin lake and cochineal red) are shown in Fig. 3.6 [186]. The responses originate from the reduction of quinone group to a diphenol. Figure 3.7 shows SWV of solid quercetin, reseda lake, carminic acid and dragon's blood, which are caused by the oxidation of hydroxyl groups in these molecules [186]. The peak potentials of these voltammograms are significantly different, which permits identification of the kind of pigment or dye.

The SWV of microparticles of the lipid-lowering drug simvastatin is shown in Fig. 3.8 [188]. The electrode reaction is totally irreversible, as indicated by the backward component of the response. The net peak potential is a linear function of the logarithm of SW frequency, as can be seen in Fig. 3.9. From the slope of this relationship (79 mV/d.u.) the product $\alpha n = 0.75$ was calculated [188]. These examples show that SWV can be used for the characterization of electrode reactions of microparticles immobilized on solid electrodes.

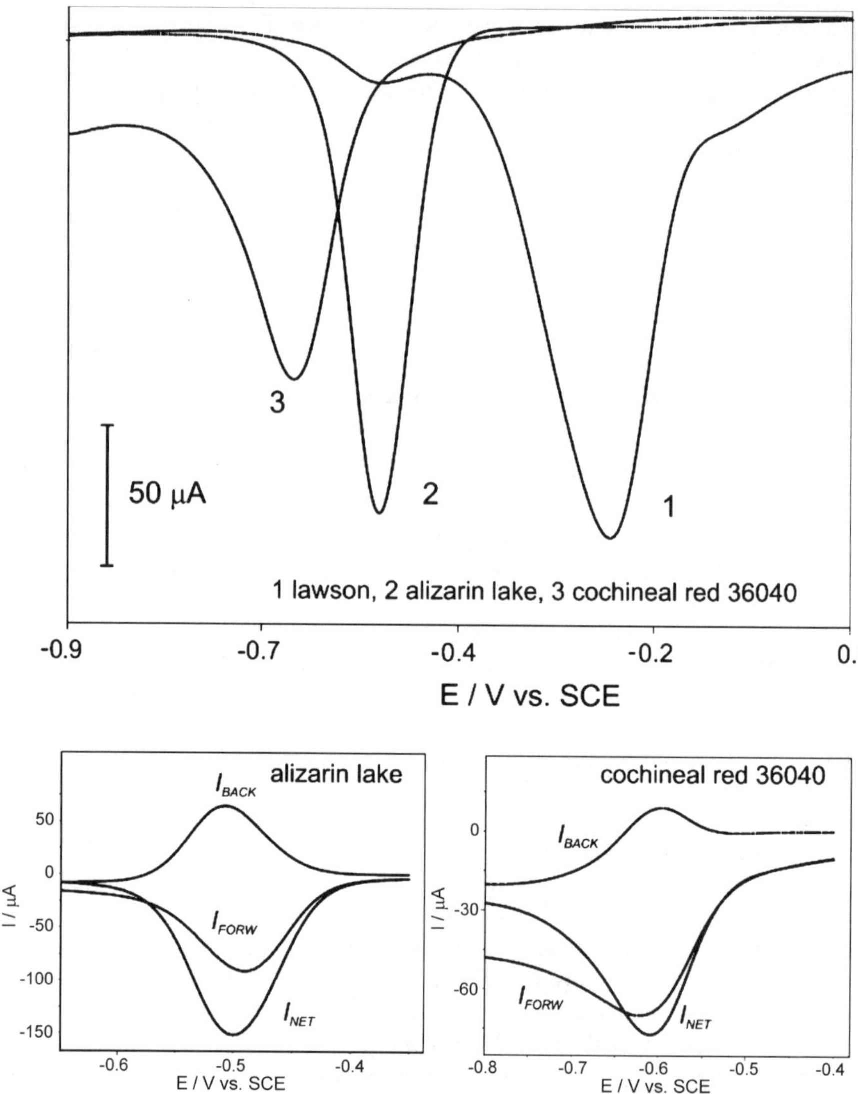

Fig. 3.6 Cathodic SWV curves for three quinone dyes and pigments: lawson (1, a quasireversible process), alizarin lake (2, a reversible process) and cochineal red (3, a quasireversible process). Scans from open-circuit potential toward negative potentials. *Insets*: the net, forward and backward current components are shown for alizarin lake and cochineal red (reprinted from [186] with permission)

3.3 Mechanistic and Kinetic Studies

SWV has been applied for the measurements of kinetic parameters of electrode reactions of adsorbed reactant and product. Standard rate constants and trans-

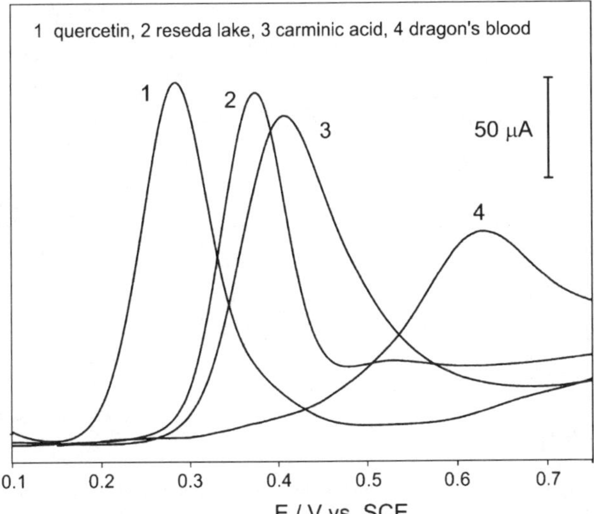

Fig. 3.7 Anodic SWV curves of flavones (1 and 2, reversible processes), an anthraquinone dye (3, a quasireversible process) and a pyran dye (4, a totally irreversible process). Scans from open-circuit potential toward positive potentials (reprinted from [186] with permission)

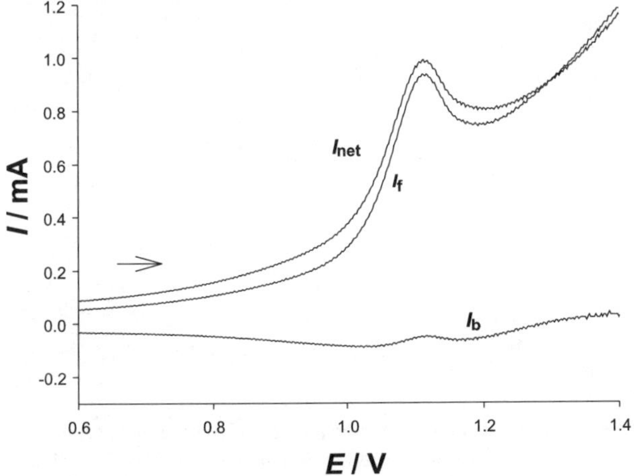

Fig. 3.8 Square-wave voltammetry of simvastatin microparticles in 0.09 M $NaClO_4$, pH 7. Net response (I_{net}) and its forward (I_f) and backward (I_b) components. Frequency is 150 Hz, amplitude is 50 mV and potential increment is 2 mV (reprinted from [188] with permission)

fer coefficients of the following adsorbed compounds and complexes were determined: adriamycin [190], alizarine red S [191], azobenzene [192,193], azurin [194], cadmium(II)-oxine [195] and cadmium(II)-ferron complexes [196], cercosporin [197], cinnoline [198,199], copper(II)-oxine complex [200], cytochrome c [201], 4-(dimethylamino)-azobenzene [202], europium(III)-salicylate complex [203], 5-fluorouracil [204], glutathione [205], 2-hydroxy-5-[(4-sulfophenyl)azo] benzoic acid

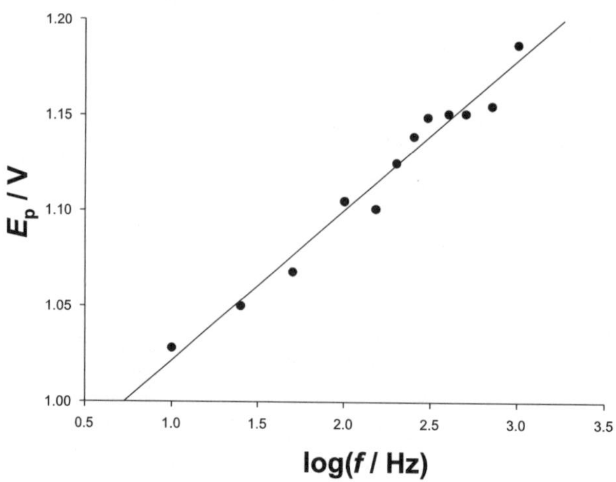

Fig. 3.9 Square-wave voltammetry of simvastatin microparticles: the dependence of peak potential on the logarithm of frequency. All other data are as in Fig. 3.8 (reprinted from [188] with permission)

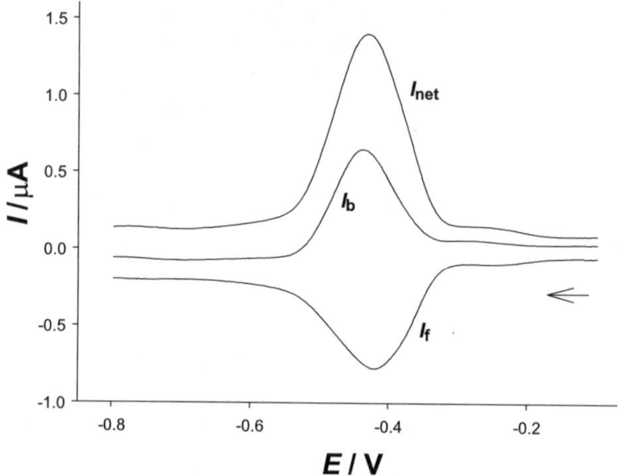

Fig. 3.10 Square-wave voltammetry of adriamycin adsorbed on mercury electrode. A net response and its forward and backward components. The concentration of adriamycin is 1.72×10^{-4} M and the supporting electrolyte is 0.9 M KNO_3, pH 4.65. Adriamycin is accumulated during 30 s from unstirred solution, at -0.1 V. $E_{sw} = 50$ mV, $f = 10$ Hz and $\Delta E = -2$ mV (reprinted from [190] with permission)

[206], indigo [207], mercury(II)-ferron complex [208], molybdenum(VI)-1,10-phenanthroline-fulvic acid complex [209, 210], molybdenum(VI)-mandelic acid complex [211], probucole [212], vanadium(V) [213] and the azo-dye Sudan III [214].

Square-wave voltammogram of antibiotic adriamycin adsorbed to the surface of static mercury drop electrode is shown in Fig. 3.10 [190]. It originates from the reduction of quinone group in the molecule. Figure 3.11 shows the dependence of the

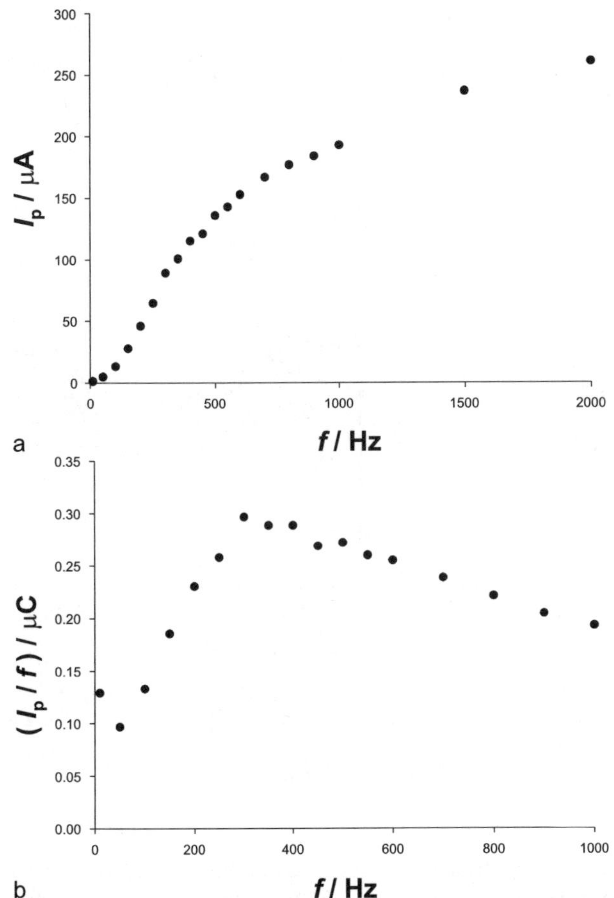

Fig. 3.11 Dependence of the net peak current of adriamycin (**a**) and the ratio between the net peak current and frequency (**b**) on the square-wave frequency. Experimental conditions are as in Figure 3.10 (reprinted from [190] with permission)

net peak current and the ratio of the net peak current and frequency on the frequency. The latter relationship is in maximum for $f_{\mathrm{max}} = 300\,\mathrm{Hz}$. Using the theoretically calculated critical kinetic parameter $\overline{\kappa}_{\mathrm{max}} = 0.49 \pm 0.12$, the standard rate constant was calculated as $k_{\mathrm{s}} = 147 \pm 36\,\mathrm{s}^{-1}$ [190].

Figure 3.12 shows the forward and backward components of square-wave voltammograms of mercury(II)-ferron complex adsorbed on the surface of static mercury drop electrode [208]. The ratio of the current and the corresponding SW frequency is reported. At pH 3.5 the electrode reaction involves the direct transfer of two electrons, whereas at pH 5.8 only one electron is exchanged. The simulated responses are presented by symbols. The best fit was achieved by using the following standard rate constants and the transfer coefficients: $k_{\mathrm{s}} = 1550 \pm 50\,\mathrm{s}^{-1}$ and $\alpha = 0.5$ (at pH 3.5), and $k_{\mathrm{s}} = 1900 \pm 400\,\mathrm{s}^{-1}$ and $\alpha = 0.55$ (at pH 5.8) [208].

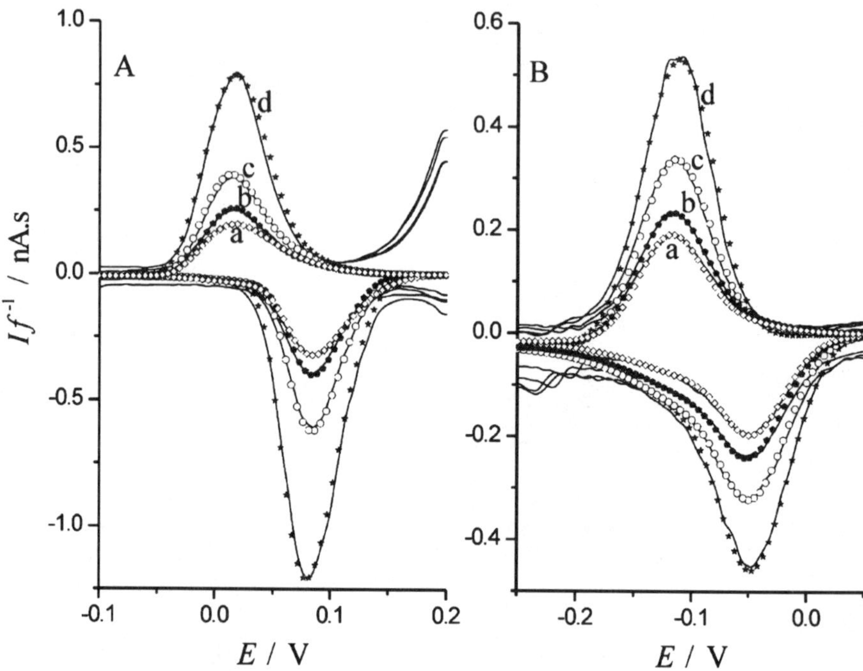

Fig. 3.12A,B Experimental (——) SW voltammograms for 0.1 M KNO$_3$ + 0.01 M H$_3$PO$_4$ + 0.55 μM ferron at pH 3.5 (**A**) and 5.8 (**B**). *Symbols* are simulated profiles calculated with $n = 2$ (**A**) and $n = 1$ (**B**). $E_{sw} = 50$ mV, $\Delta E = 5$ mV, $E_{ini} = E_{acc} = 0.2$ V, $t_{acc} = 20$ s and $f/$Hz $= 300$ (*a*), 400 (*b*), 500 (*c*) and 700 (*d*) (reprinted from [208] with permission)

SWV was used for the investigation of charge transfer kinetics of dissolved zinc(II) ions [215–218] and uranyl-acetylacetone [219], cadmium(II)-NTA [220] and ruthenium(III)-EDTA complexes [221], and the mechanisms of electrode reactions of bismuth(III) [222], europium(III) [223, 224] and indium(III) ions [225], 8-oxoguanine [226] and selenium(IV) ions [227, 228]. It was also used for the speciation of zinc(II) [229, 230], cadmium(II) and lead(II) ions in various matrices [231–235].

The enhancement of SWV net peak current caused by the reactant adsorption on the working electrode surface was utilized for detection of chloride, bromide and iodide induced adsorption of bismuth(III), cadmium(II) and lead(II) ions on mercury electrodes [236–243]. An example is shown in Fig. 3.13. The SWV net peak currents of lead(II) ions in bromide media are enhanced in the range of bromide concentrations in which the neutral complex PbBr$_2$ is formed in the solution [239]. If the simple electrode reaction is electrochemically reversible, the net peak current is independent of the composition of supporting electrolyte. So, its enhancement is an indication that one of the complex species is adsorbed at the electrode surface.

The diminished net peak currents of dissolved metal ions in solutions of electroinactive, surface-active substances were used for the quantitative analysis of de-

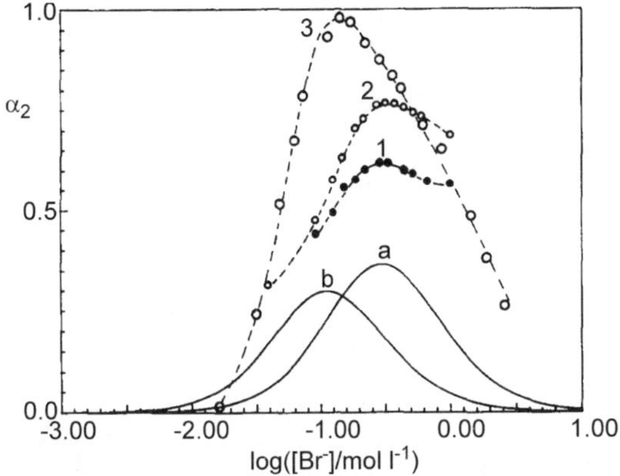

Fig. 3.13 Normalized DPP (1) and SWV (2 and 3) peak currents of lead in bromide media at ionic strengths $= 1$ (1, 2) and 4 (3) mol/l in comparison with the percentage of $PbBr_2$ (a, b) in the solution. pH 2; total lead concentration: 5×10^{-5} M. Cumulative stability constants: $\log \beta_1 = 1.10$, $\log \beta_2 = 1.8$, $\log \beta_3 = 2.2$, $\log \beta_4 = 2.0$ (i.s. $= 1$ M); $\log \beta_1 = 1.48$, $\log \beta_2 = 2.5$, $\log \beta_3 = 3.5$, $\log \beta_4 = 3.5$, $\log \beta_5 = 2.7$ (i.s. $= 4$ M) (reprinted from [239] with permission)

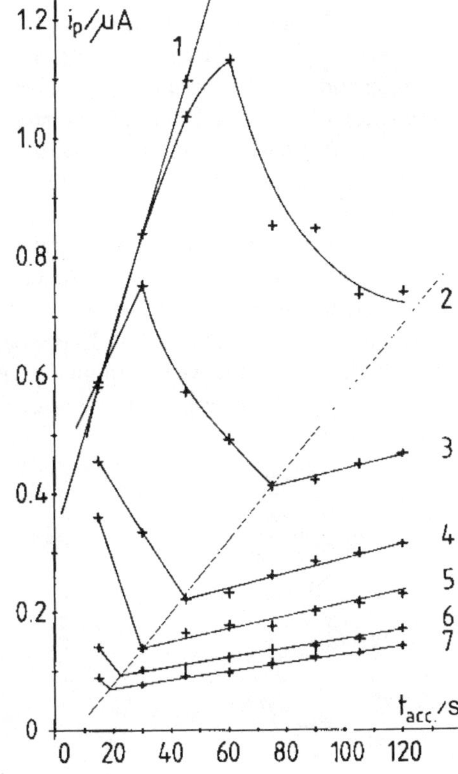

Fig. 3.14 Anodic stripping square-wave voltammetry (ASSWV) of 1×10^{-6} M Cd^{2+}. Dependence of peak currents on the accumulation time. $E_{acc} = -0.8$ V, $t_{acc} = 5$ s. (1) organic carbon-free water and (2) double-distilled water contaminated by unknown surfactant. Additions of Triton X-100 to (2) in mg/l: (3) 0.1, (4) 0.3, (5) 0.5, (6) 0.8 and (7) 1. $E_{sw} = 30$ mV, $f = 100$ Hz and $\Delta E = 2.4$ mV (reprinted from [247] with permission)

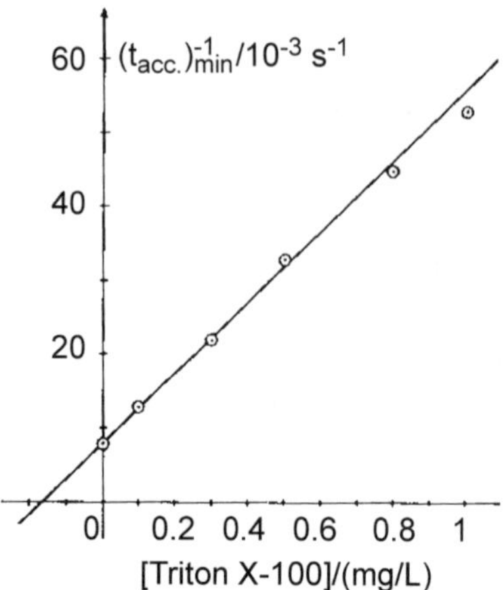

Fig. 3.15 ASSWV of Cd^{2+}. Dependence of inverse values of accumulation times which correspond to the minima of (I_p vs. t_{acc}) curves (shown in Fig. 3.14) on the standard additions of Triton X-100. Experimental conditions as in Fig. 3.14 (reprinted from [247] with permission)

tergents [244–247]. This is shown in Fig. 3.14. The SWV anodic stripping net peak currents of Cd^{2+} on static mercury drop electrode depend on the duration of accumulation and the concentration of detergent Triton X-100 [247]. The adsorptive accumulation of Triton X-100 occurs during the electrodeposition of cadmium amalgam. The relationship between the peak currents and the accumulation time may exhibit a maximum and a minimum. The adsorbed Triton X-100 decreases the efficacy of amalgam accumulation by blocking the electrode surface, and obstructs the subsequent reoxidation by diminishing the rate of charge transfer. After the minimum, the redox reaction proceeds on the electrode surface totally covered by the adsorbed surfactant. Figure 3.15 shows that the time required to attain the minimum is inversely proportional to Triton X-100 concentration [247]. This relationship can be used for the quantitative determination of surface-active substances.

Finally, SWV was applied to monitor the progress of homogeneous chemical reactions [248, 249].

References

1. Cabral MF, de Souza D, Alves CR, Machado SAS (2003) Eclet Quim 28:41
2. Beckett EL, Lawrence NS, Evans RG, Davis J, Compton RG (2001) Talanta 54:871
3. Özkan SA, Uslu B, Sentürk Z (2004) Electroanalysis 16:231
4. Barra CM, Correia dos Santos MM (2001) Electroanalysis 13:1098
5. dos Santos LBO, Abate G, Masini JC (2004) Talanta 62:667
6. Vacek J, Andrysik Z, Trnkova L, Kizek R (2004) Electroanalysis 16:224
7. Komorsky-Lovrić Š, Gagić S, Penovski R (1999) Anal Chim Acta 389:219
8. Pavlova V, Mirčeski V, Komorsky-Lovrić Š, Petrovska-Jovanović S, Mitrevski B (2004) Anal Chim Acta 512:49
9. Komorsky-Lovrić Š (1990) J Electroanal Chem 289:161
10. Zen JM, Ting YS (1997) Anal Chim Acta 342:175
11. Golcu A, Dogan B, Ozkan SA (2005) Talanta 67:703
12. Seymour EH, Lawrence NS, Compton RG (2003) Electroanalysis 15:689
13. Chen SM, Chzo WY (2006) J Electroanal Chem 587:226
14. Yu JJ, Huang W, Hibbert B (1997) Electroanalysis 9:544
15. Yardimci C, Özaltin N (2004) Anal Bioanal Chem 378:495
16. Demircigil BT, Özkan SA, Coruh Ö, Yilmaz S (2002) Electroanalysis 14:122
17. Nagaosa Y (1979) Talanta 26:987
18. Ferret DJ, Milner GWC, Smales AA (1954) Analyst 79:731
19. Ferret DJ, Milner GCW (1956) Analyst 81:193
20. Luther III GW, Nuzzio DB, Wu J (1994) Anal Chim Acta 284:473
21. Ferret DJ, Milner GWC (1956) J Chem Soc 12:1186
22. Pon-Saravanan N, Venugopalan S, Senthilkumar N, Santhosh P, Kavita B, Gurumallesh-Prabu H (2006) Talanta 69:656
23. Çakir S, Atayman I, Çakir O (1997) Mikrochim Acta 126:237
24. Zen JM, Chen TL, Shih Y (1998) Anal Chim Acta 369:103
25. Xu Q, Wang SF (2005) Mikrochim Acta 151:47
26. Guertler O, Anh CX (1970) Mikrochim Acta 151:941
27. Nagaosa Y (1979) Fresenius Z Anal Chem 296:259
28. Chen JC, Chung HH, Hsu CT, Tsai DM, Kumar AS, Zen JM (2005) Sensors Actuators B 110:364
29. Uslu B, Özkan SA, Sentürk Z (2006) Anal Chim Acta 555:341
30. Agui L, Lopez-Guzman JE, Gonzalez-Cortes A, Yanez-Sedeno P, Pingarron JM (1999) Anal Chim Acta 385:241
31. Bettazzi F, Palchetti I, Sisalli S, Mascini M (2006) Anal Chim Acta 555:134
32. Uslu B, Dogan B, Özkan SA, Aboul-Enein HY (2005) Anal Chim Acta 552:127
33. Lima JLFC, Loo DV, Delerue-Matos C, Roque da Silva AS (1999) Farmaco 54:145
34. Hernandez SR, Ribero GG, Goicoechea HC (2003) Talanta 61:743

35. von Sturm F, Ressel M (1962) Z Anal Chem 186:63
36. Buchanan EB, Schroeder TD, Novosel B (1970) Anal Chim Acta 42:370
37. Buchanan Jr EB, Soleta DD (1983) Talanta 30:459
38. Wojciechowski M, Go W, Osteryoung J (1985) Anal Chem 57:155
39. Wechter C, Sleszynski N, O'Dea JJ, Osteryoung J (1985) Anal Chim Acta 175:45
40. Ostapczuk P, Valenta P, Nürnberg HW (1986) J Electroanal Chem 214:51
41. Kumar V, Heineman WR (1987) Anal Chem 59:842
42. Zamponi S, Berrettoni M, Marassi R (1989) Anal Chim Acta 219:153
43. Wojciechowski M, Balcerzak J (1990) Anal Chem 62:1325
44. Wang J, Tian B (1992) Anal Chem 64:1706
45. Wang J, Brennsteiner A, Angnes L, Sylwester A, LaGasse RR, Bitsch N (1992) Anal Chem 64:151
46. Wang J, Rongrong X, Baomin T, Wang J, Renschler CL, White CA (1994) Anal Chim Acta 293:43
47. Brett CMA, Garcia BQ, Lima JLFC (1995) Portugaliae Electrochim Acta 13:505
48. Keller OC, Buffle J (2000) Anal Chem 72:943
49. Malakhova NA, Miroshnikova EG, Stojko NY, Brainina HZ (2004) Anal Chim Acta 516:49
50. Locatelli C (2004) Electroanalysis 16:1478
51. Carapuça HM, Monterroso SCC, Rocha LS, Duarte AC (2004) Talanta 64:566
52. Golas J, Osteryoung J (1986) Anal Chim Acta 181:211
53. Golas J, Osteryoung J (1986) Anal Chim Acta 186:1
54. Wojciechowski M, Balcerzak J (1990) Anal Chim Acta 237:127
55. Wojciechowski M, Balcerzak J (1991) Anal Chim Acta 249:433
56. Nyholm L, Wikmark G (1993) Anal Chim Acta 273:41
57. De Vitre RR, Tercier ML, Tsacopoulos M, Buffle J (1991) Anal Chim Acta 249:419
58. Komorsky-Lovrić Š (1988) Anal Chim Acta 204:161
59. Banks CE, Hyde ME, Tomčik P, Jacobs R, Compton RG (2004) Talanta 62:279
60. Demetriades D, Economou A, Voulgaropoulos A (2004) Anal Chim Acta 519:167
61. Hočevar SB, Švancara I, Vytras K, Ogorevc B (2005) Electrochim Acta 51:706
62. Hardcastle JL, Murcott GG, Compton RG (2000) Electroanalysis 12:559
63. Davis J, Cardosi MF, Brown I, Hetheridge MJ, Compton RG (2001) Anal Lett 34:2375
64. Hardcastle JL, Compton RG (2002) Electroanalysis 14:753
65. Riley JP, Wallace GG (1991) Electroanalysis 3:191
66. Charalambous A, Economou A (2005) Anal Chim Acta 547:53
67. da Silva SM (1998) Electroanalysis 10:722
68. Saterlay AJ, Agra-Gutierrez C, Taylor MP, Marken F, Compton RG (1999) Electroanalysis 11:1083
69. Crowley K, Cassidy J (2002) Electroanalysis 14:1077
70. Tomčik P, Banks CE, Compton RG (2003) Electroanalysis 15:1661
71. Richter EM, Pedrotti JJ, Angnes L (2003) Electroanalysis 15:1871
72. Wang J, Tian B (1993) Anal Chim Acta 274:1
73. Turyan I, Mandler D (1994) Electroanalysis 6:838
74. Zen JM, Chung MJ (1995) Anal Chem 67:3571
75. Walcarius A, Devoy J, Bessiere J (2000) J Solid State Electrochem 4:330
76. Davis J, Vaughan DH, Stirling D, Nei L, Compton RG (2002) Talanta 57:1045
77. Spano N, Panzanelli A, Piu PC, Pilo MI, Sanna G, Seeber R, Tapparo A (2005) Anal Chim Acta 553:201
78. Kruusma J, Banks CE, Nei L, Compton RG (2004) Anal Chim Acta 510:85
79. Hernandez P, Dabrio-Ramos M, Paton F, Ballesteros Y, Hernandez L (1997) Talanta 44:1783
80. Oliveira-Brett AM, Piedade JAP, Chiorcea AM (2002) J Electroanal Chem 538-539:267
81. Gazy AAK (2004) Talanta 62:575
82. Farghaly OAEM, Mohamed NAL (2004) Talanta 62:531
83. Xu G, O'Dea JJ, Mahoney LA, Osteryoung JG (1994) Anal Chem 66:808
84. Nigović B, Komorsky-Lovrić Š, Šimunić B (2005) Electroanalysis 17:839

85. Correia dos Santos MM, Famila V, Simoes Gonçalves ML (2002) Anal Bioanal Chem 374:1074
86. Komorsky-Lovrić Š (1987) J Electroanal Chem 219:281
87. Komorsky-Lovrić Š (2000) Electroanalysis 12:599
88. Ioannides X, Economou A, Voulgaropoulos A (2003) J Pharm Biomed Anal 33:309
89. Parham H, Zargar B (2005) Talanta 65:776
90. El-Desoky HS, Ghoneim EM, Ghoneim MM (2005) J Pharm Biomed Anal 39:1051
91. Radi A, Wahdan T, Abd El-Ghany N (2003) J Pharm Biomed Anal 31:1041
92. Marchiando NC, Zon MA, Fernandez H (2005) Anal Chim Acta 550:199
93. Webber A, Shah M, Osteryoung J (1983) Anal Chim Acta 154:105
94. Pacheco WF, Farias PAM, Aucelio RQ (2005) Anal Chim Acta 549:67
95. Alghamdi AH, Belal FF, Al-Omar MA (2006) J Pharm Biomed Anal 41:989
96. Hernandez P, Ballesteros Y, Galan F, Hernandez L (1994) Electroanalysis 6:51
97. Fdez de Betoño S, Arranz Garcia A, Arranz Valentin JF (1999) J Pharm Biomed Anal 20:621
98. Ly SY, Kim YH, Han IK, Moon IG, Jung WW, Jung SY, Sin HJ, Hong TK, Kim MH (2006) Microchem J 82:113
99. Sun YX, Wang SF, Zhang XH, Huang YF (2006) Sensors Actuators B 113:156
100. Ghoneim EM, El-Desoky HS, Ghoneim MM (2006) J Pharm Biomed Anal 40:255
101. Zhao C, Pan Y, He C, Guo Z, Sun L (2003) Anal Sci 19:607
102. Mirčeski V, Jordanoski B, Komorsky-Lovrić Š (1998) Portugaliae Electrochim Acta 16:43
103. Roque da Silva AMS, Lima JC, Oliva Teles MT, Oliveira Brett AM (1999) Talanta 49:611
104. Nouws HPA, Delerue-Matos C, Barros AA, Rodrigues JA, Santos-Silva A (2005) Anal Bioanal Chem 382:1662
105. Ly SY, Kim YK (2006) Sensors Actuators A 127:41
106. Castro AA, Wagener ALR, Farias PAM, Bastos MB (2004) Anal Chim Acta 521:201
107. Guiberteau A, Galeano T, Mora N, Parrilla P, Salinas F (2001) Talanta 53:943
108. Burgoa Calvo ME, Dominguez Renedo O, Arcos Martìnez MJ (2005) Anal Chim Acta 549:74
109. Radi A, Beltagi AM, Ghoneim MM (2001) Talanta 54:283
110. Beltagi AM, Khashaba PY, Ghoneim MM (2003) Electroanalysis 15:1121
111. Radi AE, Ghoneim M, Beltagi A (2001) Chem Pharm Bull 49:1257
112. Arranz A, Villalba MF, Fdez de Betoño S, Moreda JM, Arranz JF (1997) Fresenius J Anal Chem 357:768
113. O'Dea JJ, Ribes A, Osteryoung JG (1993) J Electroanal Chem 345:287
114. Rodriguez J, Berzas JJ, Castañeda G, Rodriguez N (2004) Electroanalysis 16:661
115. Trindade MAG, da Silva GM, Ferreira VS. (2005) Microchem J 81:209
116. Webber A, Shah M, Osteryoung J (1984) Anal Chim Acta 157:1
117. Webber A, Osteryoung J (1984) Anal Chim Acta 157:17
118. Özaltin N, Yardimci C, Süslü I (2002) J Pharm Biomed Anal 30:573
119. Pedrosa V de A, Codognoto L, Avaca LA (2003) J Braz Chem Soc 14:530
120. Ghoneim MM, El-Ries M, Hassanein AM, Abd-Elaziz AM (2006) J Pharm Biomed Anal 41:1268
121. Ghoneim MM, Radi A, Beltagi AM (2001) J Pharm Biomed Anal 25:205
122. De Souza D, Machado SAS (2005) Anal Chim Acta 546:85
123. De Souza D, Machado SAS, Pires RC (2006) Talanta 69:1200
124. Beltagi AM (2003) J Pharm Biomed Anal 31:1079
125. Codognoto L, Machado SAS, Avaca LA (2002) Diamond Relat Mater 11:1670
126. Radi A (2003) J Pharm Biomed Anal 33:687
127. Rodriguez J, Berzas JJ, Castañeda G, Rodriguez N (2004) Talanta 62:427
128. Guzman-Vazquez de Prada A, Reviejo AJ, Pingarron JM (2006) J Pharm Biomed Anal 40:281
129. Berzas JJ, Rodriguez J, Castañeda G (1997) Anal Chim Acta 349:303
130. Abd El-Hady D, Abdel-Hamid MI, Seliem MM, Andrisano V, Abo El-Maali N (2004) J Pharm Biomed Anal 34:879
131. Harnandez P, Galan-Estella F, Hernandez L (1992) Electroanalysis 4:45

132. Gazy AA, Mahgoub H, Khamis EF, Youssef RM, El-Sayed MA (2006) J Pharm Biomed Anal 41:1157
133. Ghoneim MM, Khashaba PY, Beltagi AM (2002) J Pharm Biomed Anal 27:235
134. Ghoneim MM, Tawfik A (2004) Anal Chim Acta 511:63
135. Temerk YM, Kamal MM, Ahmed GAW, Ibrahim HSM (2003) Anal Sci 19:1115
136. Kefala G, Economou A, Sofoniou M (2006) Talanta 68:1013
137. Wang J, Thongngamdee S, Lu D (2006) Anal Chim Acta 564:248
138. Hajian R, Shams E (2003) Anal Chim Acta 491:63
139. Arancibia V, Alarcon L, Segura R (2004) Anal Chim Acta 502:189
140. Yantasee W, Lin Y, Fryxell GE, Busche BJ (2004) Anal Chim Acta 502:207
141. Paneli M, Voulgaropoulos AV, Kalcher K (1993) Mikrochim Acta 110:205
142. Bobrowski A, Bond AM (1991) Electroanalysis 3:157
143. Bobrowski A, Bond AM (1992) Electroanalysis 4:975
144. Yarnitzky C, Schreiber-Stanger R (1986) J Electroanal Chem 214:65
145. Galeano Diaz T, Guiberteau A, Lopez Soto MD, Ortiz JM (2006) Food Chem 96:156
146. Mlakar M, Branica M (1991) Anal Chim Acta 247:89
147. Mlakar M (1992) Anal Chim Acta 260:51
148. Li H, Smart RB (1997) J Electroanal Chem 429:169
149. Economou A, Fielden PR (1993) Anal Chim Acta 273:27
150. Morfobos M, Economou A, Voulgaropoulos A (2004) Anal Chim Acta 519:57
151. Legeai S, Bois S, Vittori O (2006) J Electroanal Chem 591:93
152. Locatelli C (2006) Anal Chim Acta 557:70
153. Weber G, Messerschmidt J (2005) Anal Chim Acta 545:166
154. Gawrys M, Golimowski J (2001) Anal Chim Acta 427:55
155. Mlakar M (1993) Anal Chim Acta 276:367
156. El-Maali NA, El-Hady DA (1999) Electroanalysis 11:201
157. Wang J, Lu D, Thongngamdee S, Lin Y, Sadik OA (2006) Talanta 69:914
158. Moretto LM, Chevalet J, Mazzocchin GA, Ugo P (2001) J Electroanal Chem 498:117
159. Ugo P, Moretto LM, Rudello D, Birriel E, Chevalet J (2001) Electroanalysis 13:661
160. Carrington NA, Yong L, Xue ZL (2006) Anal Chim Acta 572:17
161. Cernak J, Blazej A, Suty L, Gigac J (1979) Cellul Chem Technol 13:621
162. Lovrić M, Pižeta I, Komorsky-Lovrić Š (1992) Electroanalysis 4:327
163. Berzas JJ, Rodriguez J, Lemus JM, Castañeda G (1993) Anal Chim Acta 273:369
164. Jordanoski B, Mirčeski V, Stojanova K (1994) Anal Lab 3:255
165. Jordanoski B, Mirčeski V, Stojanova K (1996) Croat Chem Acta 69:37
166. Stone CG, Cardosi MF, Davis J (2003) Anal Chim Acta 491:203
167. Prasad PVA, Arunachalam J, Gangadharan S (1994) Electroanalysis 6:589
168. Mirčeski V, Petrovska-Jovanović S, Stojanova K, Jordanoski B (1994) J Serb Chem Soc 59:573
169. Sladkov V, Bolyos A, David F (2002) Electroanalysis 14:128
170. Bertolino FA, Torriero AAJ, Salinas E, Olsina R, Martinez LD, Raba J (2006) Anal Chim Acta 572:32
171. Luther III GW, Branson Swartz C, Ullman WJ (1988) Anal Chem 60:1721
172. Wong GTF, Zhang LS (1992) Talanta 39:355
173. Parham H, Zargar B (2002) Anal Chim Acta 464:115
174. Arancibia V, Lopez A, Zuñiga MC, Segura R (2006) Talanta 68:1567
175. Komorsky-Lovrić Š, Lovrić M, Bond AM (1992) Anal Chim Acta 258:299
176. Komorsky-Lovrić Š (1995) J Electroanal Chem 397:211
177. Komorsky-Lovrić Š, Bartoll J, Stößer R, Scholz F (1997) Croat Chem Acta 70:563
178. Komorsky-Lovrić Š (1998) Croat Chem Acta 71:263
179. Domenech-Carbo A, Domenech-Carbo MT, Osete-Cortina L, Gimeno-Adelantado JV, Sanchez-Ramos S, Bosch-Reig F (2003) Electroanalysis 15:1465
180. Domenech-Carbo A, Sanchez-Ramos S, Yusa-Marco DJ, Moya-Moreno M, Gimeno-Adelantado JV, Bosch-Reig F (2004) Anal Chim Acta 501:103

181. Domenech-Carbo A, Sanchez-Ramos S, Yusa-Marco DJ, Moya-Moreno M, Gimeno-Adelantado JV, Bosch-Reig F (2004) Electroanalysis 16:1814
182. Komorsky-Lovrić Š, Horvat AJM, Ivanković D (2006) Croat Chem Acta 79:33
183. Komorsky-Lovrić Š, Mirčeski V, Scholz F (1999) Mikrochim Acta 132:67
184. Komorsky-Lovrić Š, Galić I, Penovski R (1999) Electroanalysis 11:120
185. Komorsky-Lovrić Š, Vukašinović N, Penovski R (2003) Electroanalysis 15:544
186. Grygar T, Kučkova Š, Hradil D, Hradilova D (2003) J Solid State Electrochem 7:706
187. Komorsky-Lovrić Š, Nigović B (2004) J Pharm Biomed Anal 36:81
188. Komorsky-Lovrić Š, Nigović B (2006) J Electroanal Chem 539:125
189. Scholz F, Schröder U, Gulaboski R (2005) Electrochemistry of immobilized particles and droplets. Springer, Berlin Heidelberg New York, 290 pp
190. Komorsky-Lovrić Š (2006) Bioelectrochem 69:82
191. Komorsky-Lovrić Š (1996) Fresenius J Anal Chem 356:306
192. O'Dea JJ, Osteryoung JG (1993) Anal Chem 65:3090
193. Komorsky-Lovrić Š, Lovrić M (1995) Electrochim Acta 40:1781
194. Zhang J, Guo SX, Bond AM, Honeychurch MJ, Oldham KB (2005) J Phys Chem B 109:8935
195. Garay F, Solis VM (2001) J Electroanal Chem 505:109
196. Garay F, Solis VM (2004) Electroanalysis 16:450
197. Marchiando NC, Zon MA, Fernandez H (2003) Electroanalysis 15:40
198. Komorsky-Lovrić Š (2002) Anal Bioanal Chem 373:777
199. Komorsky-Lovrić Š (2002) Electroanalysis 14:888
200. Garay F, Solis V, Lovrić M (1999) J Electroanal Chem 478:17
201. Reeves JH, Song S, Bowden EF (1993) Anal Chem 65:683
202. O'Dea JJ, Osteryoung JG (1997) Anal Chem 69:650
203. Lovrić M, Mlakar M (1995) Electroanalysis 7:1121
204. Mirčeski V, Gulaboski R, Jordanoski B, Komorsky-Lovrić Š (2000) J Electroanal Chem 490:37
205. Mladenov M, Mirčeski V, Gjorgoski I, Jordanoski B (2004) Bioelectrochem 65:69
206. Komorsky-Lovrić Š, Nigović B (2005) Croat Chem Acta 78:85
207. Komorsky-Lovrić Š (2000) J Electroanal Chem 482:222
208. Garay F (2003) J Electroanal Chem 548:11
209. Quentel F, Mirčeski V, Laouenan A, Elleouet C, Madec CL (2003) Electroanalysis 15:270
210. Quentel F, Mirčeski V (2003) Electroanalysis 15:1787
211. Quentel F, Mirčeski V (2004) Electroanalysis 16:1690
212. Mirčeski V, Lovrić M, Jordanoski B (1999) Electroanalysis 11:660
213. Mirčeski V, Gulaboski R, Petrovska-Jovanović S, Stojanova K (2001) Portugaliae Electrochim Acta 19:25
214. Mirčeski V, Gulaboski R (2003) Croat Chem Acta 76:37
215. Tamamushi R, Matsuda K (1977) J Electroanal Chem 80:201
216. Jindal HL, Matsuda K, Tamamushi R (1978) J Electroanal Chem 90:185
217. Jindal HL, Matsuda K, Tamamushi R (1978) J Electroanal Chem 90:197
218. Matsuda K, Tamamushi R (1979) J Electroanal Chem 100:831
219. Ćosović B, Jeftić Lj, Branica M, Galus Z (1973) Croat Chem Acta 45:475
220. Raspor B, Branica M (1975) J Electroanal Chem 59:99
221. Prakash R, Tyagi B, Ramachandraiah G (1997) Indian J Chem A 36:201
222. Komorsky-Lovrić Š, Lovrić M. Branica M (1993) J Electrochem Soc 140:1850
223. Kinard WF, Philp Jr RH (1970) J Electroanal Chem 25:373
224. Zelić M (2003) Croat Chem Acta 76:241
225. Zelić M, Mlakar M, Branica M (1994) Anal Chim Acta 289:299
226. Oliveira Brett AM, Piedade JAP, Serrano SHP (2000) Electroanalysis 12:969
227. Zelić M, Branica M (1990) Electroanalysis 2:452
228. Zelić M (2005) Electroanalysis 17:1302
229. Komorsky-Lovrić Š, Lovrić M, Branica M (1986) J Electroanal Chem 214:37
230. Komorsky-Lovrić Š, Branica M (1987) J Electroanal Chem 226:253
231. Branica M, Pižeta I, Branica-Jurković G, Zelić M (1989) Mar Chem 28:227

232. Zelić M, Branica M (1992) Anal Chim Acta 268:275
233. Zelić M, Branica M (1992) Electroanalysis 4:701
234. Chakrabarti CL, Cheng JG, Lee WF, Back MH, Schroeder WH (1996) Environ Sci Technol 30:1245
235. Lam MT, Murimboh J, Hassan NM, Chakrabarti CL (2001) Electroanalysis 13:94
236. Lovrić M, Branica M (1986) J Electroanal Chem 214:103
237. Komorsky-Lovrić Š, Lovrić M, Branica M (1988) J Electroanal Chem 241:329
238. Komorsky-Lovrić Š, Lovrić M, Branica M (1989) J Electroanal Chem 266:185
239. Zelić M, Branica M (1991) J Electroanal Chem 309:227
240. Zelić M, Branica M (1992) Electroanalysis 4:623
241. Komorsky-Lovrić Š, Branica M (1993) J Electroanal Chem 358:273
242. Zelić M, Branica M (1992) Anal Chim Acta 262:129
243. Zelić M, Pižeta I (1997) Electroanalysis 9:155
244. Komorsky-Lovrić Š, Branica M (1993) Anal Chim Acta 276:361
245. Plavšić M, Krznarić D, Ćosović B (1994) Electroanalysis 6:469
246. Komorsky-Lovrić Š, Branica M (1994) Fresenius J Anal Chem 349:633
247. Komorsky-Lovrić Š, Lovrić M, Branica M (1995) Electroanalysis 7:652
248. Macfie G, Atherton JH, Compton RG (2002) Electroanalysis 14:479
249. Bragato C, Daniele S, Baldo MA (2005) Electroanalysis 17:1370

Chapter 4
Square-Wave Voltammetry at Liquid–Liquid Interface

Charge transfer phenomena across an interface between two immiscible liquids (most frequently, water and an organic solvent) have been a subject of permanent interest in electrochemistry over the past decades. Liquid interfaces (L–L) are of exceptional importance for biomimetic studies of living cell membranes, separation membranes, ion selective electrodes, solvent extraction, etc. In a major part of electrochemical studies concerned with the interface between two immiscible electrolyte solutions (ITIES), a four-electrode experimental configuration has been used, combined with cyclic voltammetry, impedance methods, and seldom, pulse voltammetric techniques. Unfortunately, in a very few studies SWV has been chosen as a voltammetric method to inspect the ion transfer reactions at the ITIES [1–3]. The pioneering application of SWV to study the ion transfer reactions at the L–L interface is related with the development of the three-phase electrodes [4, 5]. This electrode system emerged as a powerful experimental tool to asses the ion transfer reactions across liquid interfaces using a conventional three-electrode assembly [6–24]. The importance of the three-phase electrodes stems from their experimental simplicity and wide applicability to a large variety of ions and liquid interfaces. A separate monograph is devoted to detailed description and application of three-phase electrodes [25]. The application of SWV for kinetic measurements of ion transfer reactions across L–L interface was conducted at thin organic film-modified electrodes [26–30]. The following chapter gives a brief account on applications of SWV to measure the thermodynamics and kinetics of ion transfer reactions across L–L interface by using three-phase and thin-film electrodes.

4.1 Three-Phase Electrodes and Their Application to Measure the Energy of Ion Transfer Across Liquid–Liquid Interface

Three-phase electrodes have been constructed in two major configurations. Most frequently, it consists of a paraffin-impregnated graphite electrode (GE) modified with a macroscopic droplet of a water immiscible organic solvent (O) (e.g., nitroben-

V. Mirčeski, Š. Komorsky-Lovrić, M. Lovrić, *Square-Wave Voltammetry*
doi: 10.1007/978-3-540-73740-7, ©Springer 2008

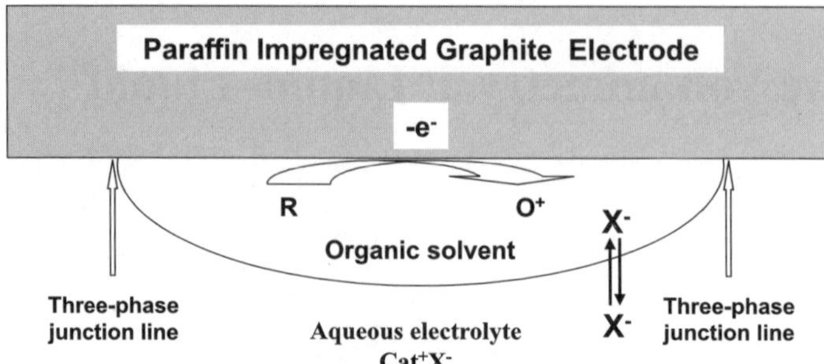

Fig. 4.1 Scheme of the three-phase electrode with a droplet configuration, consisting of a paraffin-impregnated graphite electrode, modified with a macroscopic droplet of an organic solvent that contains a neutral redox probe

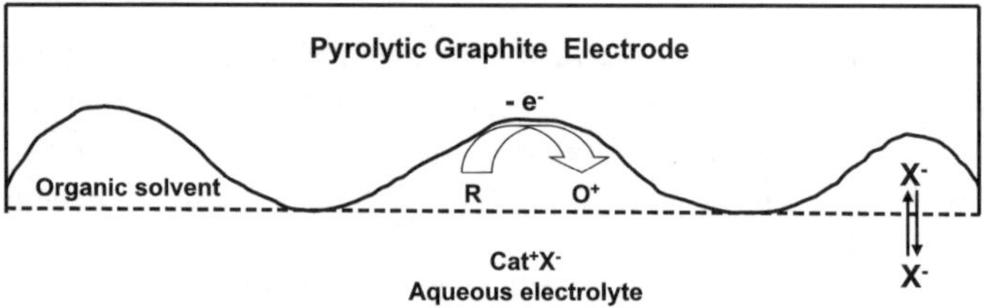

Fig. 4.2 Scheme of a part of the three-phase electrode consisting of pyrolytic graphite electrode modified with an uneven thin film of an organic solvent covering partly the electrode surface and containing a neutral redox probe

zene (nb)) (Fig. 4.1) [4, 5]. In the second configuration (Fig. 4.2) an edge plane pyrolytic graphite electrode is used modified with an uneven thin film of the organic solvent covering partly the electrode surface [21, 23]. Following the immersion of the electrode into the aqueous electrolyte (W) an interface is formed between the two liquid phases. The connecting line where the electrode, the aqueous solution, and the organic solvent meet together is termed as a three-phase boundary line (Fig. 4.1). The organic phase contains only a neutral lipophilic redox probe (e.g., decamethylferrocene (DMFC)), without including any deliberately added electrolyte. Upon electrochemical transformation of the redox probe at the GE–O interface a charge excess is produced in the organic phase, which is simultaneously compensated by ingress of corresponding ions from the aqueous to the organic phase. Hence, the overall electrochemical process at the three-phase electrode couples electron and ion transfer reactions taking place at different interfaces. An oxidative mechanism coupled with anion transfer is given by the following overall electrode reaction:

$$R_{(O)} + X^-_{(W)} \rightleftarrows O^+_{(O)} + X^-_{(O)} + e^- \tag{4.1}$$

Figure 4.3 shows a representative voltammogram recorded at a three-phase electrode with a droplet configuration consisting of DMFC as a redox probe and nitrobenzene as the organic solvent. The oxidation of DMFC to decamethylferrocenium cation

(DMFC$^+$) is coupled to ingress of SCN$^-$ from water to the organic phase. In spite of the fact that the organic phase does not contain any added supporting electrolyte, voltammetric curves are well developed. The electrochemical reaction (4.1) commences along the three-phase boundary line, where the electrode serves as a source or a sink for electrons, the organic phase provides the redox active material and the aqueous phase supplies the charge compensating ions, and, most importantly, where already at the beginning of the experiment the resistance within the organic phase will be sufficiently low to ensure a potential drop at the graphite-organic phase interface that can drive the oxidation of DMFC [31]. A detailed analysis of the properties of the response showed that the electrochemical reaction is mainly confined to the three-phase boundary region, thus showing features of an electrode process occurring in a restricted diffusion space [32]. For these reasons, the model described in Sect. 2.7 can be used for a qualitative description of the voltammetric properties of three-phase electrodes. Taking into account that the electrical conductivity of the organic phase is provided only by the partition of the aqueous electrolyte, the model has been extended by introducing the effect of an uncompensated resistance [20]. It has been shown that the resistance effect depends on a complex resistance parameter defined as $\rho = R_\Omega \frac{n^2 F^2}{RT} A\, c_R^* \sqrt{Df}$, where R_Ω is the resistance of the organic phase, and the other symbols have their usual meaning. As the resistance effect resembles the charge transfer kinetics, criteria to distinguish these two phenomena have been developed [20]. The net peak current and potential vary in a similar way with both resistance and kinetics of the reaction. However, the half-peak width behaves considerably different, being significantly increased due to the uncompensated resistance, while being almost insensitive to the kinetics of the electrochemical reaction. This is the basis for distinguishing between the two effects. A detail analysis of the response recorded at the two types of three-phase electrodes when nitrobenzene was used as an organic solvent [23] showed that the system is not affected significantly by the uncompensated resistance for moderate frequencies ranging up to 300 Hz.

The application of three-phase electrodes for determining the energy of ion transfer is based on precise measurements of the formal potential of reaction (4.1). The latter is defined as [4, 5]:

$$E_c^{\theta'} = E_{O_{(O)}^+|R_{(O)}}^{\theta} + \Delta_W^O \varphi_{X^-}^{\theta} - \frac{RT}{F}\ln\left(a_{X_{(W)}^-}^*\right) + \frac{RT}{F}\ln\left(\frac{a_{R_{(O)}}^*}{2}\right) \tag{4.2}$$

where $E_{O_{(O)}^+|R_{(O)}}^{\theta}$ is the standard potential of the redox couple in the organic phase, $\Delta_W^O \varphi_{X^-}^{\theta}$ is the standard potential of the transfer of the anions X$^-$ from water to the organic phase, $a_{R_{(O)}}^*$ is the activity of the redox probe in the bulk of the organic phase, and $a_{X_{(W)}^-}^*$ is the activity of the X$^-$ in the bulk of the aqueous phase. Equation (4.2) is valid only for the condition $a_{R_{(O)}}^* \ll a_{X_{(W)}^-}^*$ [4]. Knowing the standard redox potential $E_{O_{(O)}^+|R_{(O)}}^{\theta}$ and measuring the formal potential $E_c^{\theta'}$, the standard potential of ion transfer $\Delta_W^O \varphi_{X^-}^{\theta}$ can be estimated. In a general case, the standard potential of transfer of an ion i with the charge number z is related with the standard Gibbs energy of

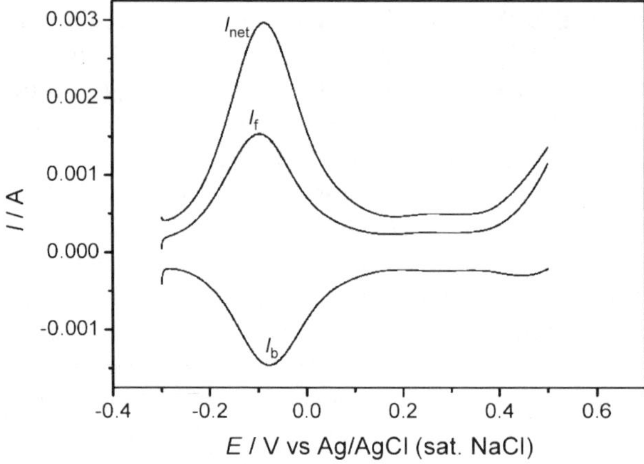

Fig. 4.3 Forward (I_f), backward (I_b) and net (I_net) components of the voltammetric response recorded at a three-phase electrode with a droplet configuration consisting of a paraffin impregnated electrode and a nitrobenzene solution of DMFC at 0.1 mol/L concentration. The electrode is immersed in 1 mol/L aqueous solution containing SCN^- anions. The other experimental conditions are: $f = 100\,\mathrm{Hz}$, $E_\mathrm{sw} = 50\,\mathrm{mV}$, and $\Delta E = 0.15\,\mathrm{mV}$ (reprint from [7] with permission)

transfer by $\Delta_\mathrm{W}^\mathrm{O} \varphi_{i^z}^\theta = -\frac{\Delta_\mathrm{W}^\mathrm{O} G_{i^z}^\theta}{zF}$, as well as with the partition coefficient of the ion by $P_{i^z} = \exp\left(\frac{zF}{RT}\Delta_\mathrm{W}^\mathrm{O}\varphi_{i^z}^\theta\right)$.

Equation (4.2) predicts a linear dependence of $E_\mathrm{c}^{\theta'}$ vs. $\Delta_\mathrm{W}^\mathrm{O}\varphi_{\mathrm{X}^-}^\theta$ with a slope 1, and a linear dependence between $E_\mathrm{c}^{\theta'}$ vs. $\log(a_{\mathrm{X}_\mathrm{w}^-}^*)$ with a slope $2.303\frac{RT}{F}$. These two dependencies can serve as diagnostic criteria to identify the electrochemical mechanism (4.1). Figure 4.4a shows the effect of different anions on the position of the net peak recorded at the three-phase electrode with a droplet configuration, where DMFC is the redox probe and nitrobenzene is the organic solvent. Figure 4.4b shows the linear variation of the net peak potential with $\Delta_\mathrm{W}^\mathrm{O}\varphi_{\mathrm{X}^-}^\theta$, with a slope close to 1. Recalling that the net peak potential of a reversible reaction is equivalent to the formal potential of the electrochemical reaction (Sect. 2.1.1), the results in Fig. 4.4 confirm the validity and applicability of Eq. (4.2).

Three-phase electrodes with a droplet configuration have been extensively used to measure the transfer energy from water to nitrobenzene of ions such as: Cl^-, NO_3^-, SCN^-, ClO_4^- [4], I^- [8], ClO_3^-, BrO_3^-, IO_4^-, OCN^-, $SeCN^-$, CN^-, N_3^- [11], K^+, Rb^+, Tl^+, Cs^+, $(CH_3)_4N^+$, $(C_2H_5)_4N^+$, $(C_3H_7)_4N^+$, $(C_6H_{11})_4N^+$, $(C_7H_{13})_4N^+$, and $(C_8H_{17})_4N^+$ [17]. Besides the transfer of inorganic ions, the transfers of a series of anionic forms of amino acids [12], peptides [16], aliphatic and aromatic mono- and dicarboxylic acids, and phenols [7] have been assessed. Particularly important is the application of the three-phase electrodes to measure the ion transfer energy across water–n-octanol interface [9, 18], as n-octanol is the standard solvent for determination of lipophilicity of compounds. Before three-phase electrodes, this solvent was inaccessible for electrochemical techniques based on the four-electrode arrangements, due to the narrow polarization window. Moreover, us-

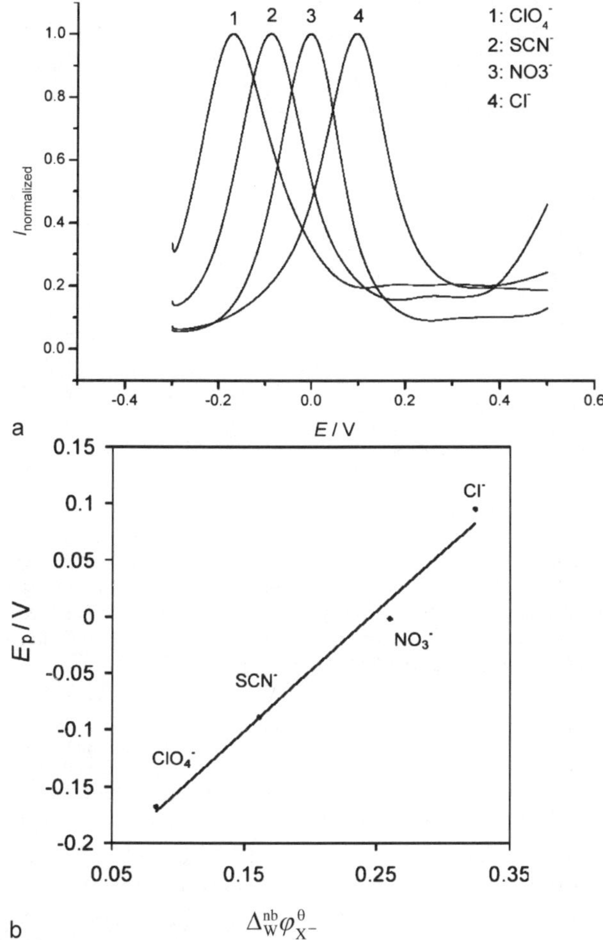

Fig. 4.4 a Normalized net voltammograms recorded at the three-phase electrode with a droplet configuration immersed in 1 mol/L aqueous solutions of different sodium salts. The organic phase is composed of 0.1 mol/L DMFC solution in nitrobenzene. The normalization is performed with respect to the peak current. The conditions are: $f = 100\,\mathrm{Hz}$, $E_{sw} = 50\,\mathrm{mV}$, and $\Delta E = 0.15\,\mathrm{mV}$. **b** The dependence of the net peak potential on the standard potential of ion transfer (reprints from [25] and [7] with permission)

ing D- and L-2-octanol [22] and menthol [13] as chiral organic solvents, the transfer of chiral anions has been studied and the effect of chirality on the energy of the ion transfer reaction has been quantified for the first time.

The three-phase electrode with a thin-film configuration (Fig. 4.2) has been mainly used in combination with nitrobenzene as an organic solvent and lutetium bis(tetra-t-butylphthalocyaninato) complexes as a redox probe (LBPC) [21,23]. Figure 4.5 depicts a typical voltammogram recorded with this redox probe in contact with 0.1 mol/L aqueous solution of KNO$_3$. LBPC can be both oxidized and re-

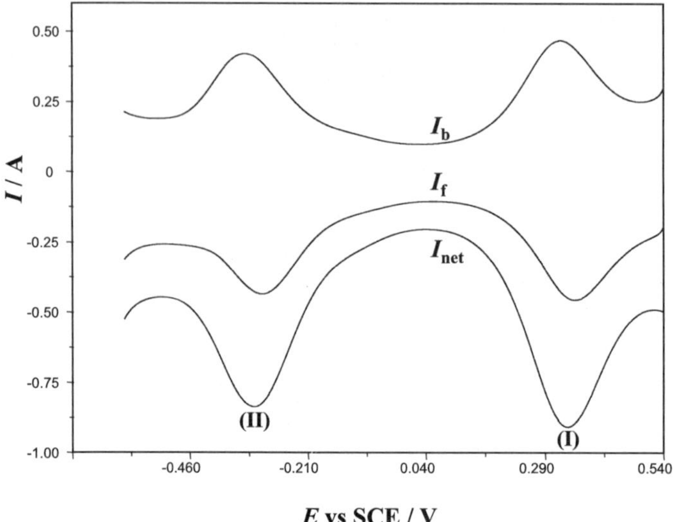

Fig. 4.5 Forward (I_f), backward (I_b) and net (I_{net}) components of the voltammetric response recorded at a three-phase electrode with a thin-film configuration consisting of an edge plane pyrolytic graphite electrode modified with a 2 mmol/L nitrobenzene solution of LBPC. The electrode is immersed in 0.5 mol L^{-1} aqueous solution of KNO$_3$. The other experimental conditions are: $f = 100$ Hz, $E_{sw} = 50$ mV, $\Delta E = 0.15$ mV (reprint from [21] with permission)

duced to a stable hydrophobic monovalent cation and anion, respectively. This enables inspection of the anion and cation transfer from water to the organic phase with one and the same redox probe. Accordingly, the response in Fig. 4.5 located at more positive potentials corresponds to the oxidation of LBPC to LBPC$^+$ accompanied by a simultaneous transfer of NO$_3^-$ from water to nitrobenzene, whereas the response at more negative potentials reflects the reduction of LBPC coupled to the K$^+$ transfer. A detail comparative study of the performances of the three-phase electrodes shown in Figs. 4.1 and 4.2 indicated that LBPC is a superior redox probe compared to DMFC [4, 5] or other redox probes [17], due to its particular chemical stability and hydrophobicity [21, 23]. Using this redox probes, the transfer of Li$^+$ and Na$^+$ ions from water to nitrobenzene without using any facilitating agents has been demonstrated for the first time. The accessible range of Gibbs energies of cation and anion transfers from water to nitrobenzene using LBPC are of about $\Delta_w^{nb} G_{\mathrm{Cat}^+}^\theta \leq 43$ kJ/mol and $\Delta_w^{nb} G_{\mathrm{X}^-}^\theta \leq 50$ kJ/mol, respectively. This interval corresponds to a potential window of about 1 V!

4.2 Analyzing the Kinetics of the Ion Transfer Across Liquid–Liquid Interface with Thin-Film Electrodes

The methodology for analyzing the kinetics of ion transfer reactions across L–L interface has been worked out for using thin-film electrodes [26–29]. The electrode assembly, shown in Fig. 4.6, is similar to the three-phase electrode with a thin-film configuration (Fig. 4.2). The thin-film electrode consists of a pyrolytic graphite electrode covered completely with a uniform film of an organic solvent layer of some micrometer thickness. The organic film contains the redox probe and a suitable electrolyte. The transferring ion, accompanying the electrochemical transformation of the redox probe, is present in a large excess in both the aqueous and organic phases. Thus, the transferring ion is at the same time the ion controlling the potential drop at the L–L interface. Following the immersion of the electrode into the aqueous electrolyte, no three-phase boundary line is formed, being the main difference with the three-phase electrode and a thin-film configuration. Similar to the three-phase electrodes, the overall electrochemical reaction (4.3) at the thin-film electrode couples the electron transfer reaction at the graphite–organic solvent interface (4.4) with the ion transfer at the organic solvent–water interface (4.5).

$$R_{(O)} + X^-_{(W)} \rightleftarrows O^+_{(O)} + X^-_{(O)} + e^- \qquad \text{(overall reaction)} \qquad (4.3)$$

$$R_{(O)} \rightleftarrows O^+_{(O)} + e^- \qquad \text{(GE–O)} \qquad (4.4)$$

$$X^-_{(W)} \rightleftarrows X^-_{(O)} \qquad \text{(O–W)} \qquad (4.5)$$

As the electrochemical reaction is confined to the boundaries of the thin film, the voltammetric response exhibits a quasireversible maximum. The position of the quasireversible maximum on the log frequency axis depends on the kinetics of the overall reaction at the thin-film electrode, i.e., reflecting the coupled electron–ion transfer (4.3). Analyzing the evolution of the quasireversible maximum measured with different redox probes and various transferring ions, it has been demonstrated

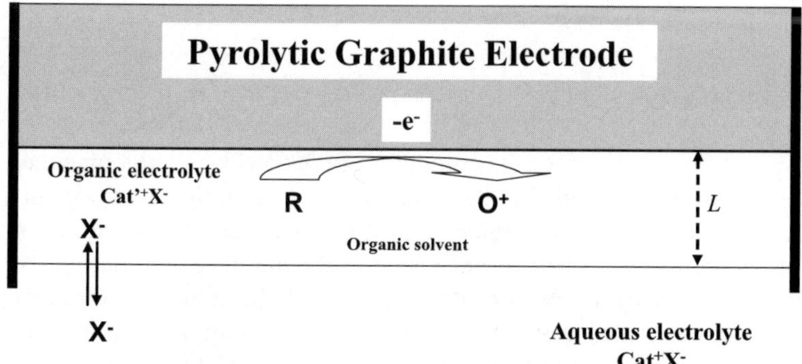

Fig. 4.6 Thin-film electrode consisting of a pyrolytic graphite electrode covered with a film of an organic solvent containing a neutral hydrophobic redox probe and a suitable electrolyte

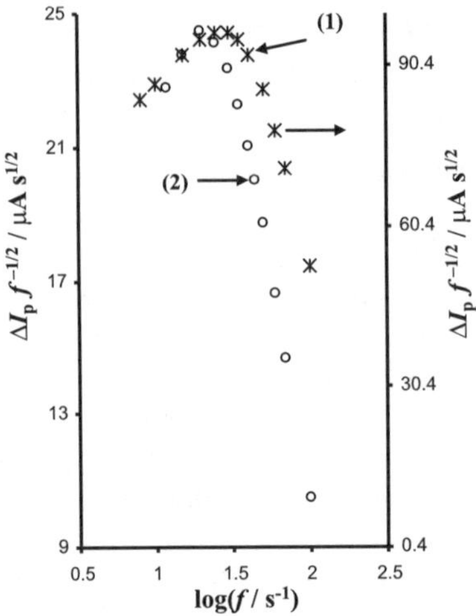

Fig. 4.7 Dependence of the ratio $\frac{\Delta I_p}{\sqrt{f}}$ on the $\log(f)$ for oxidation of LBPC (*curve 1*) and DMFC (*curve 2*) measured in 0.1 mol/L aqueous solution of LiClO$_4$. The electrode was covered with 1 μL nitrobenzene solution containing 0.1 mol/L (C$_4$H$_9$)$_4$NClO$_4$ and 2 mmol/L of the corresponding redox compound. The other experimental conditions were: $E_{sw} = 50$ mV, $\Delta E = 0.15$ mV (with permission from [27])

that the ion transfer reaction is the rate controlling step [26]. This is illustrated by the results shown in Figs. 4.7 and 4.8. Figure 4.7 shows that the position of the quasi-reversible maximum is independent on the redox probe for the transfer of ClO$_4^-$ driven by either oxidation of DMFC or LBPC. On the other hand, using a single redox probe and different transferring ions, the position of the maximum is significantly different (Fig. 4.8). These results allow concluding that the rate of the overall reaction (4.3) is controlled by the ion transfer across the L–L interface of the thin-film electrode.

In the theoretical modeling, the kinetics of anion transfer is assumed to obey the Butler-Volmer equation [29]:

$$\frac{I}{FA} = k_{s,it}\exp(\beta_{it}\varphi_{O|W})\left[\left(c_{X^-_{(W)}}\right)_{x=L} - \exp(-\varphi_{O|W})\left(c_{X^-_{(O)}}\right)_{x=L}\right] \tag{4.6}$$

Here, $(c_{X_{(W)}})_{x=L}$ and $(c_{X_{(O)}})_{x=L}$ are concentrations of the transferring ion on either side of the L–L interface positioned at the distance $x = L$ from the electrode surface, where L is the thickness of the organic film (Fig. 4.6). Here the difference of distances between the location at the organic side of the interface and that on the aqueous side of the interface has been neglected. β_{it} and $k_{s,it}$ are the transfer coefficient and standard rate constant of the ion transfer, respectively. The ion transfer is driven by the potential difference at the O–W interface, $\Delta_W^O\varphi$. The potential difference at the O–W interface is related with the potential at the GE–O interface and the potential difference between the electrode and the aqueous phase GE–W that is

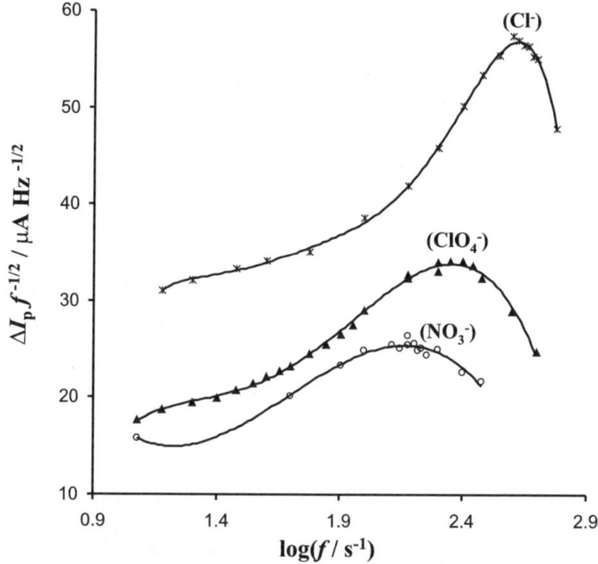

Fig. 4.8 Effect of the type of the aqueous ions on the position of the quasireversible maxima of LBPC. The concentrations of the aqueous solutions are 1 mol/L. In the case of perchlorate the cation is Li$^+$, and for both nitrate and chloride it is K$^+$. All other conditions are the same as in the caption of Fig. 4.7 (with permission from [27])

potentiostatically controlled. Hence, the following equation holds:

$$\varphi_{\text{GE}|\text{O}} + \varphi_{\text{O}|\text{W}} = \varphi_{\text{GE}|\text{W}} \tag{4.7}$$

Here, $\varphi_{\text{GE}|\text{O}} = \frac{F}{RT}(E_{\text{GE}|\text{O}} - E^{\theta}_{\text{O}|\text{R}})$, $\varphi_{\text{GE}|\text{W}} = \frac{F}{RT}(E_{\text{GE}|\text{W}} - E^{\theta'}_{\text{c}})$, where $E^{\theta'}_{\text{c}} = E^{\theta}_{\text{O}|\text{R}} + \Delta^{\text{O}}_{\text{W}}\varphi^{\theta}_{\text{X}}$ is the formal potential of reaction (4.3) and $E^{\theta}_{\text{O}|\text{R}}$ is the standard potential of the redox probe in the organic phase.

At the electrode surface the following condition holds:

$$(c_{\text{O}})_{x=0} = (c_{\text{R}})_{x=0}\exp(\varphi_{\text{GE}|\text{O}}) \tag{4.8}$$

Furthermore, the conversion rate of the redox couple at the GE–O is controlled by the current, i.e.,

$$\left(\frac{\partial c_{\text{R}}}{\partial x}\right)_{x=0} = -\left(\frac{\partial c_{\text{O}}}{\partial x}\right)_{x=0} = \frac{I}{FAD} \tag{4.9}$$

The physical meaning of (4.9) and (4.6) is that the two charge transfer processes taking place at different interfaces must proceed at equal rate. Combining (4.6),

(4.7), and (4.8), one obtains the following equation:

$$\frac{I}{FA} = k_{s,it}\exp(\beta_{it}\varphi_{GE|W})\left[\frac{(c_O)_{x=0}}{(c_R)_{x=0}}\right]^{-\beta_{it}}$$

$$\left\{\left(c_{X_{(W)}^-}\right)_{x=L} - \exp(-\varphi_{GE|W})\left[\frac{(c_O)_{x=0}}{(c_R)_{x=0}}\right]\left(c_{X_{(O)}^-}\right)_{x=L}\right\} \tag{4.10}$$

As the transferring ion is present in a large excess in both liquid phases compared to the redox probe, its concentrations at the O–W are virtually constant and equal to their concentrations in the bulk of the liquid phases. Thus, (4.10) is transformed into the following form:

$$\frac{I}{FA} = k_{s,it}c^*_{X_{(W)}^-}\exp(\beta_{it}\varphi_{GE|W})\left[\frac{(c_O)_{x=0}}{(c_R)_{x=0}}\right]^{-\beta_{it}}\left\{1 - \exp(-\varphi_{GE|W})\left[\frac{(c_O)_{x=0}}{(c_R)_{x=0}}\right]\rho_1\right\} \tag{4.11}$$

where $\rho_1 = \dfrac{c^*_{X_{(O)}^-}}{c^*_{X_{(W)}^-}}$ is the concentration ratio. The latter reaction is the basis for the simulations of the quasireversible process at thin-film electrodes controlled by the rate of the ion transfer reaction. The other aspects of the model referring to the mass transfer regime of redox active species within the thin film are identical as described in Sect. 2.7.

The simulations revealed that the apparent reversibility of the overall reaction (4.3) depends on the kinetic parameter $\kappa = \dfrac{k_{s,it}}{\sqrt{Df}}$ and the two concentration ratios, $\rho_1 = \dfrac{c^*_{X_{(O)}^-}}{c^*_{X_{(W)}^-}}$ and $\rho_2 = \dfrac{c^*_{X_{(W)}^-}}{c^*_R}$. Figure 4.9 shows the theoretical quasireversible maxima calculated by varying κ, for different concentrations of the transferring ion in the aqueous phase. Note that the dimensionless current is defined as $\Psi = I(FAc^*_R\sqrt{Df})^{-1}$. The shift of the quasireversible maximum by changing the concentration of the transferring ion is an intrinsic property of the overall reaction controlled by the ion transfer kinetics. Variation of the concentration of any electroactive species causes changes in the observed electrochemical reversibility. For instance, an increase of $c^*_{X_{(W)}^-}$ accelerates the rate of the ion transfer, resulting in an increase of the apparent reversibility. For these reasons, the maximum is achieved for lower critical values of κ_{max} by increasing of $c^*_{X_{(W)}^-}$ (Fig. 4.9). It is worth noting that the relation $\log(\kappa_{max})$ vs. $\log(c^*_{X_{(W)}^-})$ is linear, with a slope of about -0.5, which is valid for the conditions given in Fig. 4.9. This property of the quasireversible maximum is the critical criterion to recognize that the ion transfer controls the kinetics of the overall process at the thin-film-modified electrode.

For a given κ, the apparent reversibility can be affected by changing $c^*_{X_{(W)}^-}$ or $c^*_{X_{(O)}^-}$. Figure 4.10 shows that $\Delta\Psi_p$ depends parabolically on $c^*_{X_{(W)}^-}$, which is a consequence of the quasireversible maximum. Note that the variation of $c^*_{X_{(W)}^-}$ corresponds to the simultaneous alteration of both ρ_1 and ρ_2. Therefore, in the real ex-

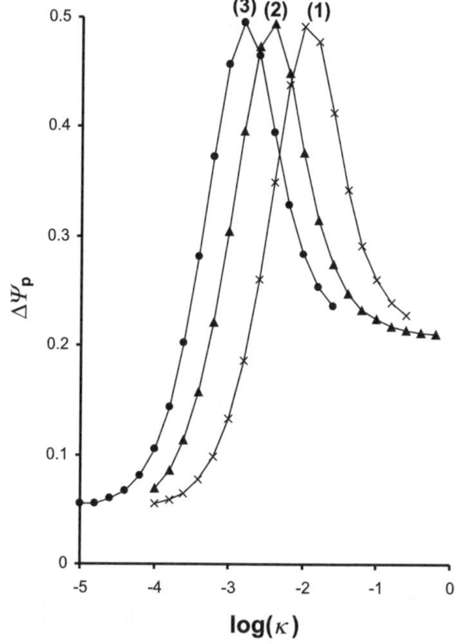

Fig. 4.9 Coupled electron–ion transfer reaction controlled by the ion transfer kinetics. Dependence of $\Delta \Psi_p$ on $\log(\kappa)$ for $c^*_{X^-_{(W)}} = 0.01$ (*1*); 0.1 (*2*), and 0.5 mol/L (*3*). The other conditions of the simulations are: $c^*_{X^-_{(O)}} = 0.1$ mol/L, $c^*_{Red} = 1$ mmol/L, $\beta_{it} = 0.5$, $\Lambda = 0.7$ ($\Lambda = L\sqrt{\frac{f}{D}}$), $E_{sw} = 50$ mV, $\Delta E = 10$ mV (with permission from [30])

periment in which the overall reaction is controlled by the ion transfer kinetics, the apparent reversibility could be varied either by adjusting f and affecting $\kappa = \frac{k_{s,it}}{\sqrt{Df}}$, or by adjusting ρ_1 and ρ_2.

The aforementioned methodology has been applied to measure the kinetics of a series of monovalent ions by using the oxidation of LBPC [26–29]. As the redox probe LBPC is oxidized to the stable hydrophobic cation LBPC$^+$, and the electrode reaction is accompanied by either anion ingress from the aqueous phase (4.12) or cation expulsion from the organic phase (4.13), which depends on the type of ions and their relative affinity for both liquid phases.

$$\text{LBPC}_{(nb)} + X^-_{(W)} \rightleftarrows \text{LBPC}^+_{(nb)} + X^-_{(nb)} + e^- \tag{4.12}$$

$$\text{LBPC}_{(nb)} + \text{Cat}^+_{(nb)} \rightleftarrows \text{LBPC}^+_{(nb)} + \text{Cat}^+_{(W)} + e^- \tag{4.13}$$

The transfer of anions (ClO$_4^-$, NO$_3^-$, SCN$^-$, Br$^-$, and Cl$^-$) and cations ((CH$_3$)$_4$N$^+$, (C$_4$H$_9$)$_4$N$^+$, Na$^+$, and K$^+$) has been studied according to mechanism (4.12) and (4.13), respectively. In all cases, the voltammetric curves were particularly well developed, with a virtually constant peak potential and half-peak width, indicating no effect of uncompensated resistance. In all experiments, the transferring ion was

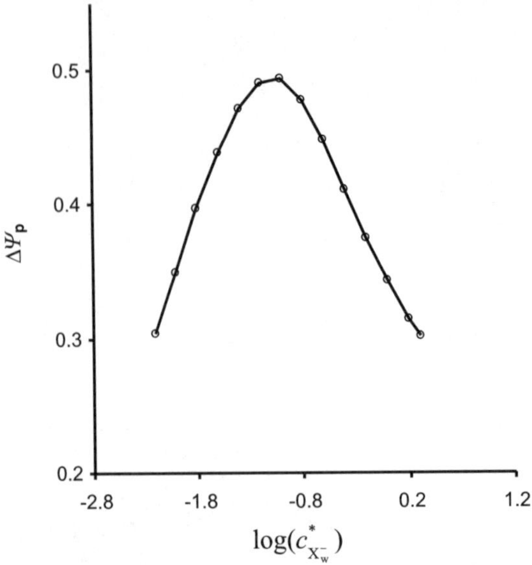

Fig. 4.10 Coupled electron–ion transfer reaction controlled by the ion transfer kinetics. Dependence of $\Delta\Psi_p$ on $\log(c^*_{X^-_{(W)}})$ for $\log(\kappa) = -2.4$. The other conditions of the simulations are the same as for Fig. 4.9 (with permission from [30])

present in both liquid phases, at concentrations at least two orders of magnitude higher than the redox probe concentration.

Figures 4.11 and 4.12 show the quasireversible maxima corresponding to the transfer of anions and cations, respectively. The position of the maximum depends on the nature of the transferring ion, confirming that the overall process is controlled by the ion transfer kinetics. For each transferring ion, the evolution of the quasireversible maximum was analyzed by varying $c^*_{X_{(W)}}$. In all cases, the position of the maximum has been shifted towards a higher critical frequency (f_{max}) by increasing $c^*_{X_{(W)}}$. This satisfies the main criterion for a control of the rate of the overall reaction by the ion transfer. The dependence $\log(f_{max}^{-1/2}/\mathrm{Hz}^{-1/2})$ vs. $\log(c^*_{X^-_{(W)}})$ for each ion is listed in Table 4.1. Comparing the position of the experimentally measured with the theoretically calculated quasireversible maxima, the rate constants have been estimated for each ion listed in Table 4.1.

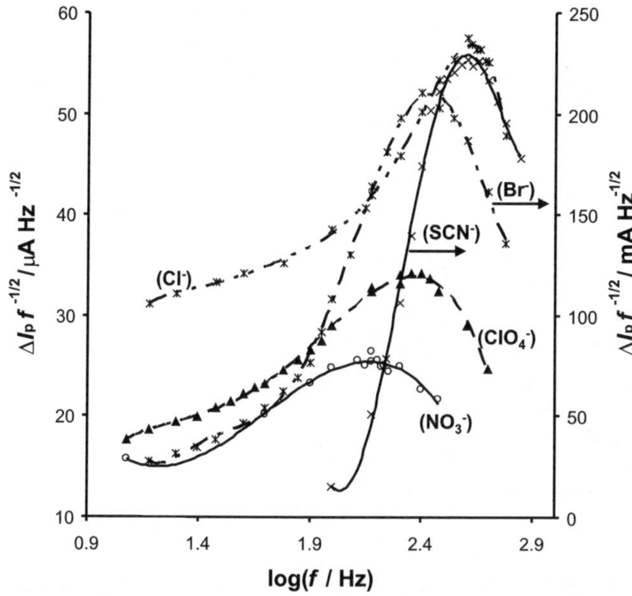

Fig. 4.11 Experimental quasireversible maxima corresponding to anion transfers from water to nitrobenzene measured by the oxidation of LBPC. The concentrations of the transferring ion are $c^*_{X^-_{(W)}} = 1$ mol/L and $c^*_{X^-_{(nb)}} = 0.1$ mol/L. All other conditions were the same as in the caption of Fig. 4.7 (with permission from [30])

Table 4.1 Analysis of the quasireversible maxima measured for three different concentrations of the transferring ion in the aqueous phase. The other experimental conditions are the same as in Fig. 4.11. The standard rate constants have been estimated from the critical frequencies measured for $c^*_{X_{(W)}} = 0.5$ mol/l

Transferring ion	$^*E_\mathrm{p}/V$	Critical frequency ($f_\mathrm{max}/$Hz) for $c^*_{X_\mathrm{w}^-}/$mol L^{-1}			$\log(f_\mathrm{max}^{-1/2}/\mathrm{Hz}^{-1/2})$ vs. $\log(c^*_{X_\mathrm{w}^-}/\mathrm{mol\,L}^{-1})$	$^*10^3 k_\mathrm{s}/\mathrm{cm\,s}^{-1}$
		0.1	0.5	1		
K^+	-0.011	60	300	550	$y = -0.467x - 1.362$	2.5
Na^+	-0.042	35	180	450	$y = -0.369x - 1.285$	1.42
$(CH_3)_4N^+$	0.244	45	200	/	$y = -0.463x - 1.290$	1.6
$(C_4H_9)_4N^+$	0.493	80	180	/	$y = -0.252x - 1.203$	1.42
Br^-	0.472	50	140	250	$y = -0.344x - 1.190$	1.06
SCN^-	0.316	70	200	400	$y = -0.369x - 1.285$	1.6
NO_3^-	0.356	20	110	150	$y = -0.454x - 1.116$	0.79
ClO_4^-	0.254	25	110	220	$y = -0.470x - 1.167$	0.79
Cl^-	0.555	35	200	400	$y = -0.531x - 1.305$	1.6

* The concentration of the transferring ion in the aqueous phase is 0.5 mol/L

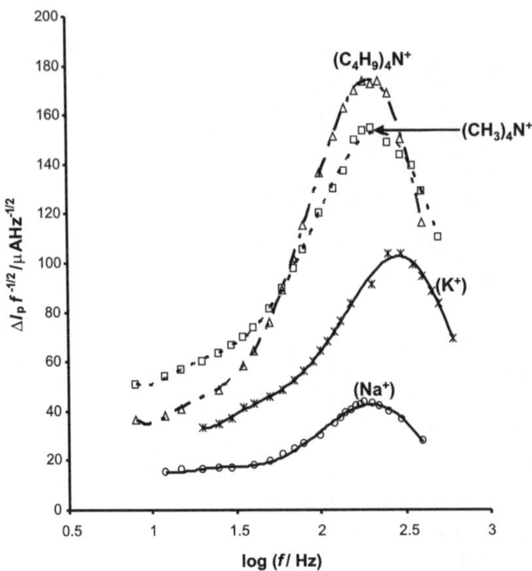

Fig. 4.12 Experimental quasireversible maxima corresponding to cation expulsion from nitrobenzene to water measured by the oxidation of LBPC. The concentrations of the transferring ion are $c^*_{X^-_{(W)}} = 0.5$ and $c^*_{X^-_{(nb)}} = 0.1$ mol/L. All other conditions are the same as in the caption of Fig. 4.7 (with permission from [30])

References

1. Beni V, Ghita M, Arrigan DWM (2005) Biosens Bioelectron 20:2097
2. Juárez AV, Baruzzi AM, Yudi LM (2005) J Electroanal Chem 577:281
3. O'Mahony AM, Scanlon MD, Berduque A, Beni V, Arrigan DWM, Faggi E, Bencini A (2005) Electrochem Commun 7:976
4. Scholz F, Komorsky-Lovrić Š, Lovrić M (2000) Electrochem Commun 2:112
5. Scholz F, Gulaboski R (2005) Chem Phys Chem 6:16
6. Komorsky-Lovrić Š, Lovrić M, Scholz F (2001) Collect Czech Chem Commun 66: 434
7. Komorsky-Lovrić Š, Riedl K, Gulaboski R, Mirčeski V, Scholz F (2002) Langmuir 18:8000 (and Komorsky-Lovrić Š, Riedl K, Gulaboski R, Mirčeski V, Scholz F (2003) Langmuir 19:3090
8. Mirčeski V, Scholz F (2002) J Electroanal Chem 522:189
9. Gulaboski R, Mirčeski V, Scholz F (2002) Electrochem Commun 4:277
10. Mirčeski V, Gulaboski R, Scholz F (2002) Electrochem Commun 4:814
11. Gulaboski R, Riedl K, Scholz F (2003) Phys Chem Chem Phys 5:1284
12. Gulaboski R, Mirčeski V, Scholz F (2003) Amino Acids 24:149
13. Scholz F, Gulaboski R, Mirčeski V, Langer P (2002) Electrochem Commun 4:659
14. Scholz F, Gulaboski R (2005) Faraday Discuss 129:169
15. Gulaboski R, Caban K, Stojek Z, Scholz F (2004) Electrochem Commun 6:215
16. Gulaboski R, Scholz F (2003) J Phys Chem B 107:5650
17. Scholz F, Gulaboski R, Caban K (2003) Electrochem Commun 5:929
18. Bouchard G, Galland A, Carrupt P-A, Gulaboski R, Mirčeski V, Scholz F, Girault H (2003) Phys Chem Chem Phys 5:3748
19. Gulaboski R, Galland A, Bouchard G, Caban K, Kretschmer A, Carrupt P-A, Stojek Z, Girault HH, Scholz F (2004) J Phys Chem B 108:4565
20. Mirčeski V, Gulaboski R, Scholz F(2004) J Electroanal Chem 566:351
21. Quentel F, Mirčeski V, L'Her M (2005) J Phys Chem B 109:1262
22. Mirceski V (2006) Electrochem Commun 8:123
23. Quentel F, Mirčeski V, L'Her M, Mladenov M, Scholz F, Elleouet C (2005) J Phys Chem B 109:13228
24. Scholz F (2006) Annu Rep Prog Chem C 102:43
25. Scholz F (2005) Schröder U, Gulaboski R (eds) Electrochemistry of immobilized particles and droplets. Springer, Berlin Heidelberg New York
26. Shi C, Anson C (1998) Anal Chem 70:3114
27. Quentel F, Mirčeski V, L'Her M (2005) Anal Chem 77:1940
28. Gulaboski R, Mirčeski V, Pereira CM, Cordeiro MNDS, Silva AF, Quentel F, L'Her M, Lovrić M (2006) Langmuir 22:3404
29. Mirčeski V, Quentel F, L'Her M, Scholz F (2006) J Electroanal Chem 586:86
30. Mirčeski V, Quentel F, L'Her M, Pondaven A (2005) Electrochem Commun 7:1122
31. Komorsky-Lovrić Š, Mirčeski V, Kabbe Ch, Scholz F (2004) J Electroanal Chem 566:371
32. Mirčeski V, Gulaboski R (2006) J Phys Chem B 110:2812

Appendix A
Mathematical Modeling of Electrode Reaction in a Thin-Layer Cell with the Modified Step-Function Method

Modeling of the electrode reaction (2.230) in a thin-layer cell has been performed by applying Laplace transforms [1] combined with the modified step-function method [2]. The original step-function method for solving linear integral equation of Volterra type, encountered frequently in modeling electrode processes, has been proposed by Nicholson and Olmstead [3], as previously described in Sect. 1.2. For complex electrode mechanisms, such as those coupled to adsorption equilibria and chemical reactions, or various electrode mechanisms in a thin-layer cell, it is very difficult, or completely impossible to give the mathematical solution in the form of an integral equation which can be further solved by the step-function method of Nicholson and Olmstead. The following part describes a procedure to overcome this limitation.

An electrode reaction in a thin-layer cell is described by the following mathematical model:

$$\frac{\partial c_R}{\partial t} = D\frac{\partial^2 c_R}{\partial x^2} \tag{A.1}$$

$$\frac{\partial c_O}{\partial t} = D\frac{\partial^2 c_O}{\partial x^2} \tag{A.2}$$

$$t = 0, \quad 0 \le x \le L: \quad c_R = c_R^*, \quad c_O = 0 \tag{A.3}$$

$$t > 0, \quad x = 0: \quad D\left(\frac{\partial c_R}{\partial x}\right) = -D\left(\frac{\partial c_O}{\partial x}\right) = \frac{I}{nFA} \tag{A.4}$$

$$t > 0, \quad x = L: \quad D\left(\frac{\partial c_R}{\partial x}\right) = -D\left(\frac{\partial c_O}{\partial x}\right) = 0 \tag{A.5}$$

Applying Laplace transformation to (A.1), one obtains:

$$s\mathscr{L}c_R - c_R^* = D\frac{\partial^2 \mathscr{L}c_R}{\partial x^2} \tag{A.6}$$

where $\mathscr{L}$ is the symbol for Laplace transform, and s is the transform variable. By introducing the substitution

$$u = \mathscr{L}c_R - \frac{c_R^*}{s}, \tag{A.7}$$

one obtains the following simplified differential equation:

$$u - \frac{D}{s}\frac{\partial^2 u}{\partial x^2} = 0 \tag{A.8}$$

with a general solution

$$u_{1/2} = C_1 e^{-\sqrt{\frac{s}{D}}x} + C_2 e^{\sqrt{\frac{s}{D}}x} \tag{A.9}$$

where C_1 and C_2 are unknown constants. Equations (A.7) and (A.9) show that the function u at $x = 0$, i.e., the electrode surface is:

$$u_{x=0} = \mathscr{L}(c_R)_{x=0} - \frac{c_R^*}{s} , \tag{A.10}$$

$$u_{x=0} = C_1 + C_2 . \tag{A.11}$$

Combining the latter two equations yields

$$\mathscr{L}(c_R)_{x=0} = \frac{c_R^*}{s} + C_1 + C_2 . \tag{A.12}$$

Hence, to find $\mathscr{L}(c_R)_{x=0}$ one needs to evaluate the constants C_1 and C_2. This can be achieved by taking into account boundary conditions (A.4) and (A.5), as described in the following part. The first derivation of (A.7) with respect to x, at $x = L$, is:

$$\left(\frac{\partial u}{\partial x}\right)_{x=L} = \left(\frac{\partial \mathscr{L} c_R}{\partial x}\right)_{x=L} . \tag{A.13}$$

Applying the Laplace transform to the boundary condition (A.5), and combining the result with (A.13), gives

$$\left(\frac{\partial u}{\partial x}\right)_{x=L} = 0 . \tag{A.14}$$

On the other hand, from (A.9), follows that

$$\left(\frac{\partial u}{\partial x}\right)_{x=L} = -\sqrt{\frac{s}{D}}C_1 e^{-\sqrt{\frac{s}{D}}L} + \sqrt{\frac{s}{D}}C_2 e^{\sqrt{\frac{s}{D}}L} . \tag{A.15}$$

Combining (A.14) and (A.15) and rearranging yields:

$$C_1 = C_2 e^{2L\sqrt{\frac{s}{D}}} \tag{A.16}$$

Substituting (A.16) in (A.9) gives:

$$u = C_2 e^{2L\sqrt{\frac{s}{D}}} e^{-\sqrt{\frac{s}{D}}x} + C_2 e^{\sqrt{\frac{s}{D}}x} \tag{A.17}$$

Applying Laplace transformation to the boundary condition (A.4), one finds:

$$\left(\frac{\partial \mathscr{L} c_R}{\partial x}\right)_{x=0} = \frac{\mathscr{L} I}{nFAD} \tag{A.18}$$

The first derivation of (A.7) and (A.17), with respect to x, at $x = 0$ is:

$$\left(\frac{\partial u}{\partial x}\right)_{x=0} = \left(\frac{\partial \mathcal{L} c_R}{\partial x}\right)_{x=0} \tag{A.19}$$

$$\left(\frac{\partial u}{\partial x}\right)_{x=0} = -C_2\sqrt{\frac{s}{D}}\,e^{2L\sqrt{\frac{s}{D}}} + C_2\sqrt{\frac{s}{D}} \tag{A.20}$$

Combining (A.18)–(A.20) gives the solution for C_2:

$$C_2 = \frac{1}{\sqrt{s}\left(1 - e^{2L\sqrt{\frac{s}{D}}}\right)}\,\mathcal{L}\,\frac{I}{nFA\sqrt{D}} \tag{A.21}$$

Substituting (A.16) and (A.21) into (A.12) one finds the solution for $\mathcal{L}(c_R)_{x=0}$

$$\mathcal{L}(c_R)_{x=0} = \frac{c_R^*}{s} + \frac{1 + e^{2L\sqrt{\frac{s}{D}}}}{\sqrt{s}\left(1 - e^{2L\sqrt{\frac{s}{D}}}\right)}\,\mathcal{L}\,\frac{I}{nFA\sqrt{D}} \tag{A.22}$$

Introducing the substitution $a = \frac{2L}{\sqrt{D}}$ and rearranging, the latter equation simplifies to the following form:

$$\mathcal{L}(c_R)_{x=0} = \frac{c_R^*}{s} + \frac{e^{-a\sqrt{s}} + 1}{\sqrt{s}\left(e^{-a\sqrt{s}} - 1\right)}\,\mathcal{L}\,\frac{I}{nFA\sqrt{D}} \tag{A.23}$$

The final solution is obtained by applying an inverse Laplace transform to (A.23), and using the convolution theorem [1]:

$$(c_R)_{x=0} = c_R^* + \frac{1}{nFA\sqrt{D}}\int_0^t I(\tau)f(t - \tau)\mathrm{d}\tau \tag{A.24}$$

Here $f(t)$ is the inverse Laplace transform of the function $F(s)$, i.e.,

$$\mathcal{L}f(t) = F(s) \tag{A.25}$$

where

$$F(s) = \frac{e^{-a\sqrt{s}} + 1}{\sqrt{s}\left(e^{-a\sqrt{s}} - 1\right)} \tag{A.26}$$

According to the step-function method of Nicholson and Olmstead, the integral equation (A.24) can be transformed into the following approximate expression:

$$(c_{R_{x=0}})_m = c_R^* + \frac{1}{nFA\sqrt{D}}\sum_{j=1}^{m} I_j \int_{(j-1)d}^{jd} f(md - \tau)\,\mathrm{d}\tau \tag{A.27}$$

The step-function method is based on incrementalization of the total time of the voltammetric experiment t by dividing it into finite equal time increments of width d, and assuming that the unknown functions, $(c_R)_{x=0}$ and $I(t)$ can be regarded as constants within each time interval. In (A.27), m is the serial number of time increments, ranging from 1 to M, where M is the total number of time increments, i.e., $Md = t$. Furthermore, $(c_{R_{x=0}})_m$ and I_m are the discrete values of the unknown functions in the time interval with the serial number m. For this procedure, the critical pre-request is to know the analytical expression of the function $f(t)$, in order to evaluate its integral over each time interval. However, for many complex electrode mechanism, the function $F(s)$ is very complex, thus obtaining the function $f(t)$ is very difficult, or completely impossible. This problem can be overcome by replacing the function $f(t)$ by an approximate expression, again evaluated by the step-function method. For this purpose, each main time increment of a width d is further divided into time sub-increments of a width $d' = \frac{d}{p}$, where p is the number of time subintervals. The serial number of each subintervals n ranges from 1 to pM. The discrete values of the function f_n will be evaluated for the each time subincrement. Thus, the next task is to obtain discrete values of the function f_n at each time subincrement.

Combining (A.25) and (A.26) yields:

$$\mathscr{L} f(t) = \frac{e^{-a\sqrt{s}} + 1}{\sqrt{s}\left(e^{-a\sqrt{s}} - 1\right)} \tag{A.28}$$

By rearranging, one obtains:

$$e^{-a\sqrt{s}} \mathscr{L} f(t) - \mathscr{L} f(t) = \frac{e^{-a\sqrt{s}}}{\sqrt{s}} + \frac{1}{\sqrt{s}} \tag{A.29}$$

The inverse Laplace transform to latter equation gives:

$$\int_0^t f(\tau) \frac{e^{-\frac{a^2}{4(t-\tau)}}}{\sqrt{\pi(t-\tau)}} \, d\tau - f(t) = \frac{e^{-\frac{a^2}{4t}}}{\sqrt{\pi t}} + \frac{1}{\sqrt{\pi t}} \tag{A.30}$$

According to the step-function method, the latter equation is replaced by the following expression:

$$\sum_{j=1}^{n} f_j M_{m-j+1} - f_n = \frac{e^{-\frac{a^2}{4nd'}}}{\sqrt{\pi nd'}} + \frac{1}{\sqrt{\pi nd'}} \tag{A.31}$$

The final formula for calculation of the discrete values of the function f_n is:

$$f_n = \frac{\dfrac{e^{-\frac{a^2}{4nd'}}}{\sqrt{\pi nd'}} + \dfrac{1}{\sqrt{\pi nd'}} - \displaystyle\sum_{j=1}^{n} f_j M_{n-j+1}}{M_1 - 1} \tag{A.32}$$

where

$$M_n = \text{erfc}\left(\frac{a}{2\sqrt{nd'}}\right) - \text{erfc}\left(\frac{a}{2\sqrt{(n-1)d'}}\right) \tag{A.33}$$

Knowing the discrete values of the function f_n over each time subincrement allows one to calculate its integral of over each main time increment, using various formulas for numerical integrations. Here, the simple quadric formula is used [4], which in a general case is defined as:

$$\int_a^b f(x)\,dx = \frac{b-a}{n} \sum_{i=1}^{n} f_i \tag{A.34}$$

Hence, the integral of the function $f(t)$ over each main time interval is calculated as follows:

$$S_m = d' \sum_{p(m-1)}^{pm} f_n \tag{A.35}$$

Finally, (A.27) gets the following form:

$$(c_{R_{x=0}})_m = c_R^* + \frac{1}{nFA\sqrt{D}} \sum_{j=1}^{m} I_j S_{m-j+1} \tag{A.36}$$

In an analogue procedure, the surface concentration of the O form is evaluated as follows:

$$(c_{R_{x=0}})_m = -\frac{1}{nFA\sqrt{D}} \sum_{j=1}^{m} I_j S_{m-j+1} \tag{A.37}$$

Combining equation (A.36), (A.37) with the Nernst equation (1.8), one obtains the final solution for a reversible electrode reaction in a thin-layer cell, given by (2.234) in Sect. 2.7.

The applicability of the foregoing procedure has been tested by modeling simple reaction under semi-infinite diffusion conditions (reaction 1.1) and EC mechanism coupled to adsorption of the redox couple (reaction (2.177)) [2]. The solutions derived by the original and modified step-function method have been compared in order to evaluate the error involved by the proposed modification. As expected, the precision of the modified step-function method depends solely on the value of p, i.e., the number of time subintervals. For instance, for the complex EC mechanism, the error was less than 2% for $p \geq 20$. This slight modification of the mathematical procedure has opened the gate toward modeling of very complex electrode mechanisms such as those coupled to adsorption equilibria and regenerative catalytic reactions [2] and various mechanisms in thin-film voltammetry [5–7].

References

1. Spiegel MR (1965) Theory and problems of Laplace transforms. McGraw-Hill, New York
2. Mirčeski V (2003) J Electroanal Chem 545:29
3. Nicholson RS, Olmstead ML (1972) Numerical solutions of integral equations. In: Matson JS, Mark HB, MacDonald HC (eds) Electrochemistry: calculations, simulations and instrumentation, vol 2. Marcel Dekker, New York, p 119
4. Deneva AP, Nančea VSD, Stojanov NV (1977) Viša Matematika, Čast 5, Državno Izdatelstvo Tehnika, Sofija 1977, p 76
5. Mirčeski V (2004) J Phys Chem B 110:2812
6. Mirčeski V, Quentel F, L'Her M, Pondaven A (2005) Electrochem Commun 7:1122
7. Mirčeski V, Quentel F, L'Her M, Elleouet C (2007) J Phys Chem C 11: 8283

Index of Symbols and Their SI Units

I. ROMAN SYMBOLS

a	Frumkin interaction parameter
A	Area, m^2
c	Concentration, $mol\,m^{-3}$
c_i^*	Bulk concentration of species i in solution, $mol\,m^{-3}$
$(c_i)_{x=0}$	Concentration of species i at the electrode surface, $mol\,m^{-3}$
c_O	Concentration of species O (oxidized form), $mol\,m^{-3}$
c_R	Concentration of species R (reduced form), $mol\,m^{-3}$
D	Diffusion coefficient, $m^2\,s^{-1}$
D_i	Diffusion coefficient of species i, $m^2\,s^{-1}$
D_O, D_R	Diffusion coefficient of species O, R, $m^2\,s^{-1}$
d	Duration of time increment, s
d'	Duration of time subincrement, s
e	Quantity of charge on the electron (elementary charge), C
ΔE	Potential increment of a staircase potential ramp
$\Delta E_{p/2}$	Half-peak width
E	Electrode potential, V
E^θ	Standard potential of electrode reaction (standard electrode potential), V
$E_c^{\theta'}$	Formal potential, V
E_p	Peak potential, V
$E_{1/2}$	Half-wave potential, V
E_{sw}	Amplitude of the SW pulses. V
f	Frequency, Hz
f_{max}	Critical frequency of quasireversible maximum, Hz
F	Faraday constant, $C\,mol^{-1}$
$\Delta_W^O G_{i^z}^\theta$	Standard Gibbs energy of the ion transfer
I	Electric current, A
I_f	Current of the forward SW component, A
I_b	Current of the backward (reverse) SW component, A
ΔI	Current of the net SW component, A

ΔI_p Net peak current, A

k_f First-order rate constant of forward reaction, s^{-1}

k_b First-order rate constant of backward reaction, s^{-1}

$k_{f,r}$ Second-order rate constant of forward reaction, $mol^{-1}\,m^3\,s^{-1}$

k_s Standard heterogeneous rate constant, $m\,s^{-1}$

k_{sur} Standard heterogeneous surface rate constant, s^{-1}

$k_{s,it}$ Standard heterogeneous rate constant of ion transfer, $m\,s^{-1}$

K Thermodynamic equilibrium constant

L Film thickness, m

m Serial number of time increments

n Serial number of time subincrements

n Number of electrons

p Number of time subincrements

pH   Negative decadic logarithm of the relative activity of H_3O^+ ions

P_i Partition coefficient

r Distance from the centre of electrode, m

r_0 Electrode radius, m

R Gas constant, $J\,mol^{-1}\,K^{-1}$

R_Ω Ohmic resistance, Ω

s Laplace transform variable

t Time, s

t_{acc} Duration of accumulation, s

t_{delay} Duration of delay period, s

t_p Duration of potential pulse, s

t_s Sampling time, s

t_{rest} Resting time, s

T Thermodynamic or absolute temperature, K

V Volume, m^3

x Distance, m

II. GREEK SYMBOLS

α	Transfer coefficient (electrochemical)
β_{it}	Ion-transfer coefficient (electrochemical)
α_a, α_c	Anodic, cathodic transfer coefficient
φ	Dimensionless electrode potential
$\Delta_W^O \varphi_{i^z}^{\theta}$	Standard potential of the ion transfer
Γ_i	Surface (excess) concentration of species i, $\mathrm{mol\,m^{-2}}$
Γ^*	Initial surface (excess) concentration, $\mathrm{mol\,m^{-2}}$
Γ_{max}	Maximal surface concentration, $\mathrm{mol\,m^{-2}}$
Θ	Surface coverage
τ	Duration of a potential cycle, s
$\mathscr{L}$	Laplace transform
Ψ	Dimensionless current
$\Psi_{p,c}$	Dimensionless cathodic peak current
$\Psi_{p,a}$	Dimensionless anodic peak current
$\Delta\Psi_p$	Dimensionless net peak current
δ	Thickness of diffusion layer

About the Authors

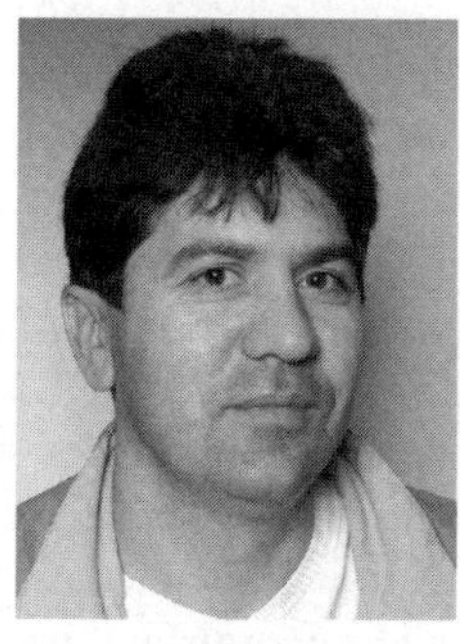

Valentin Mirčeski

Valentin Mirčeski is Associate Professor at the University "Ss Cyril and Methodius" in Skopje, in the Republic of Macedonia. He earned his PhD at the University of Zagreb in Croatia (with Dr. Milivoj Lovrić). As a Humboldt research fellow he carried out a posdoc research stay at the University of Greifswald, Germany (with Prof. F. Scholz). He is the author of more than 70 research papers, most of them concerned with the methodological and theoretical development of square-wave voltammetry. He is a contributing author of the "Electrochemical Dictionary" (Springer, in press). In 2006, his research achievements have been recognized by the French Chemical Society by awarding him the "Young Investigator Award" at the International Conference of Electroanalysis "ESEAC 2006" in Bordeaux, France.

Dr. Šebojka Komorsky-Lovrić

Dr. Šebojka Komorsky-Lovrić is senior scientist at the Department for Marine and Environmental Research, "Ruđer Bošković" Institute, Zagreb, Croatia. She obtained the PhD in chemistry at the Faculty of Pharmacy and Biochemistry, University of Zagreb (1984). She specialized in biochemistry at the Department of Chemistry, University of North Carolina, Chapel Hill, USA (Prof. L.L. Spremulli, 1985–1986) and in electroanalytical techniques at the Institute of Applied Physical Chemistry, Atomic Institute, Jülich, Germany (Prof. M. Branica, 1986–1987) and at the Department of Chemistry, Deakin University, Geelong, Australia, (Prof. A.M. Bond, 1989–1990). She was visiting scientist at the Institute of Applied Analytical and Environmental Chemistry, Humboldt University, Berlin, and at the University of Greifswald, Germany (both with Prof. F. Scholz) and at Oxford University, UK, (with Prof. R.G. Compton). Her scientific interests include the development of electroanalytical methods for the determination of organic and biochemical substances and ionic states of trace metal ions in natural waters, an investigation of the electrochemical properties of solid microparticles of inorganic salts and minerals and organic compounds mechanically attached to the surface of solid, inert electrodes and their application for the direct analyses of powders, the development of fast, simple, and inexpensive methods for the identification of the active compounds in pharmaceutical preparations and the study of electrokinetic and thermodynamic properties of organic and inorganic electrode reactions using complex voltammetric and potentiometric methods under different hydrodynamic conditions. She is the author of 100 scientific papers and participated in 33 scientific congresses with 42 presentations. She wrote two chapters in "Electroanalytical Methods"; F. Scholz (Ed.) Springer, 2002. The book has been translated into Russian (Binom, Moscow, 2006). She is a recipient of the annual award of "Zagrebačka banka" for the environmental project and, together with M. Lovrić and F. Scholz, she received the "Best Cited Paper Award 2003" from the Elsevier for a publication in *Electrochemistry Communications* on the transfer of ions across the interface of two immiscible liquids.

Milivoj Lovrić

Dr. Milivoj Lovrić is senior scientist at the Department for Marine and Environmental Research, "Ruđer Bošković" Institute, Zagreb, Croatia. He obtained the PhD in chemistry at the University of Zagreb (1983). He specialized in electrochemistry at the State University of New York, Buffalo, USA (Prof. J. Osteryoung, 1980–1981), at the University of North Carolina, Chapel Hill, USA (Prof. R.W. Murray, 1985–1986), at the Institute of Applied Physical Chemistry, Atomic Institute, Jülich, Germany (Prof. M. Branica, 1986–1987) and at Deakin University, Geelong, Australia (Prof. A.M. Bond, 1989–1990). He was visiting scientist at the Institute for Applied Analytical and Environmental Chemistry, Humboldt University, Berlin and at the Institute of Chemistry, University of Greifswald, Germany (with Prof F. Scholz). His scientific interests include theoretical electrochemistry and electroanalytical chemistry, kinetic measurements in electrochemistry, development of theories for hydrodynamic electrochemical systems, theory of electrochemical mechanisms involving reactant and product adsorption on the working electrode surface, application of theory to analytical procedures based on adsorptive accumulation of analytes and environmental electrochemistry. He published 137 scientific papers and participated in 36 scientific congresses with 47 contributions. He was mentor of two doctor thesis. He is a member of editorial boards of the *Croatica Chemica Acta* and the *Journal of Solid State Electrochemistry* and a member of the Croatian Chemical Society. He wrote two chapters in "Electroanalytical Methods", F. Scholz (Ed.), Springer, Berlin, 2002. Together with Š. Komorsky-Lovrić and F. Scholz he received the "Best Cited Paper Award 2003" from Elsevier for a publication in *Electrochemistry Communications* on the transfer of ions across the interface of two immiscible liquids, and he was awarded as "Referee of the Year 2006" of the *Journal of Solid State Electrochemistry* (Springer).

About the editor

Fritz Scholz

Fritz Scholz is Professor at the University of Greifswald, Germany. Following studies of chemistry at Humboldt University, Berlin, he obtained a Dr. rer. nat. and a Dr. sc. nat. (habilitation) from that university. In 1987 and 1989 he worked with Alan Bond in Australia. His main interest is in electrochemistry and electroanalysis. He has published more than 230 scientific papers, and he is editor and co-author of the book "Electroanalytical Methods" (Springer, 2002 and 2005, Russian Edition: BINOM, 2006), coauthor of the book "Electrochemistry of Immobilized Particles and Droplets" (Springer 2005), co-editor of the "Electrochemical Dictionary" (Springer, in press), and co-editor of volumes 7a and 7b of the "Encyclopedia of Electrochemistry" (Wiley-VCH 2006). In 1997 he founded the *Journal of Solid State Electrochemistry* (Springer) and has served as Editor-in-Chief since that time. He became editor of the series "Monographs in Electrochemistry" (Springer) in which modern topics of electrochemistry will be presented.

Index

Printing: Krips bv, Meppel, The Netherlands
Binding: Stürtz, Würzburg, Germany